Technology in America

Technology in America

A History of Individuals and Ideas

Third Edition

edited by Carroll Pursell

The MIT Press
Cambridge, Massachusetts
London, England

Library of Congress Cataloging-in-Publication Data

Names: Pursell, Carroll W., editor.
Title: Technology in America : a history of individuals and ideas / edited by
    Carroll Pursell.
Description: Third edition. | Cambridge, MA : The MIT Press, [2018] |
    Includes bibliographical references and index.
Identifiers: LCCN 2018010200 | ISBN 9780262535779 (pbk. : alk. paper)
Subjects: LCSH: Technology--United States--History.
Classification: LCC T21 .T43 2018 | DDC 609.73--dc23 LC record available at
https://lccn.loc.gov/2018010200
10 9 8 7 6 5 4 3 2 1

For my brother, Tim Pursell

Contents

Preface to the Third Edition

All fields of knowledge grow organically, and over time reveal not only new facts but also new subjects of interest, new methodologies, and new perspectives on old topics. This third edition of *Technology in America* contains three new essays that cover topics that have gained increased interest since the last: Rachel Carson and the rise of the environmental movement, A. C. Gilbert and the development of an American toy industry along with the shaping of play, and Lewis Latimer and the struggle of African Americans to gain recognition as professional inventors and engineers. These reflect a broadening interest in the field to look not only at the iconic tools, machines, and techniques of production, transportation, and communication but also ways in which technologies lent meanings to work, play, and the environment.

The founders of the history of technology field wrote many of the chapters in the original edition of this collection; eight of them are by onetime presidents of the Society for the History of Technology. These essays were inevitably written out of the scholarly consensus that prevailed a generation ago, as understood by the pioneers of the discipline. In this sense, they have become themselves important benchmarks of our evolving understandings of technology as well as being still-solid narratives of fact. The student of today would do well to ask in what ways the field has matured beyond these stories—and perhaps what may have been neglected even in more recent efforts to understand that large and powerful, but also amorphous, subject we call American technology.

Carroll Pursell
Australian National University
Canberra, Australian Capital Territory
May, 2018

Preface to the Second Edition

During the eight years since this book was first published, it has found a useful niche in the college and university classroom. The experience of almost a decade has shown that while the book does not do everything (there is still room for a number of interpretative textbooks and certainly a collection or two of primary documents, for instance), it is a tribute to the quality of the authors' scholarship that the book continues to give a fair representation of the way in which the history of American technology is viewed.

I have taken this opportunity to add three more chapters covering the topics of scientific management, electronic communications (through the example of the loudspeaker), and the cluster of electronic devices as well as institutional arrangements that came together to create the computer industry of the Silicon Valley. It is suggestive that the book's first essay, by Brooke Hindle, and its last, by James C. Williams, both highlight the social contexts in which a large number of individuals worked to produce and improve the nation's technology. Perhaps the golden age of the American inventor lies somewhere between.

The list of books for further reading has been thoroughly revised to include titles that have appeared since 1981. There are, of course, many more that could have been added, but the interested reader can find a start here for further study.

Carroll Pursell
Case Western Reserve University
Cleveland, Ohio
June 8, 1989

Preface

The score of essays in this volume focus on the spread and elaboration of American technology, and the men and women who shaped this great human activity in its American context. The biographical organization is not intended to convey the argument that history moves forward only through the deeds of great men and women. At the same time, it is true that technology is a fundamental form of human behavior, and ultimately it is people who have the ideas, design the machines and processes, build the institutions, and suffer the costs and benefits of technological change. The people here focused on include some of the best known in the pantheon of American folk heroes, but their lives serve primarily as windows through which we may observe the interplay of ideas and institutions that combined to shape the tools of their times.

The essays divide roughly between the nineteenth and twentieth centuries. After a brief introductory essay, a chapter (properly highlighting an occupational type rather than a single individual) describes the technology of America's wooden age. This is followed by a discussion of Thomas Jefferson's explicit perception of the role of technology in a democratic society, and then the introduction of the Industrial Revolution into the young nation and the nation's fundamental contribution to that great epoch: the American system of manufactures of Eli Whitney and others. The work of Thomas P. Jones illustrates the institutionalization of this industrialization in ongoing educational forms. Cyrus Hall McCormick epitomizes the spread of the industrialization to agriculture, the garden of Jefferson's yeomen farmers, and James B. Eads allows us to explain the way in which capital and engineering combined to build a transcontinental business civilization of cities linked by transportation networks.

The career of James B. Francis introduces the beginnings of a scientific technology that has, in our own times, come more and more to crowd out traditional, empirical ways of doing things. Alexander Graham Bell's telephone was only one startling application of electricity—a science made into an industry by the hands of such as he and Thomas Alva Edison, perhaps the most important transitional figure in the history of American technology. As the story of George Eastman makes clear, the use of science to transform technologies, which can then be made and marketed by large corporations, became a salient feature of twentieth-century American culture.

Ellen Swallow Richards stands as a reminder that technology touches women in our society in many ways differently than it does men, and the domestic workplace is just that—a place where people (mostly women) work, with or without machines and industrial standards of efficiency. Just as Richards stood for the science of a controlled (domestic) environment, Gifford Pinchot pioneered this Progressive Era reform among the natural resources of the nation, giving birth to the first conservation crusade and the ideal of efficient utilization. The use of science to rationalize various environments was nowhere more dramatically displayed than in the automobile factories of Henry Ford. Harnessing scientific management to the American system of manufactures, Ford's assembly line became the hallmark of the modern industrial age just as his Model T automobile became the representative artifact of modern American civilization.

The tensions inherent in trying to revolutionize American life in the name of traditional values, to guarantee political democracy with an increasingly authoritarian technology, were dramatically symbolized by the first transatlantic airplane flight accomplished by Charles Lindbergh. It was not lost on everyone that the Lone Eagle was entirely dependent on the very rational and structured industrial order that his flight was thought to transcend. In their own ways Buster Keaton and Charlie Chaplin, classic heroes of Hollywood's golden age, also helped the nation confront what it meant to live in the modern technological age.

The development of cheap, flexible, divisible energy in the form of electricity was a consistent thread running through the career of Morris L. Cooke: a goal that promised to bring the blessings of modern technology within the reach of all Americans. But already Enrico Fermi, through his work on nuclear power, and Robert H. Goddard, with his pioneer experiments in rocket flight, were pointing the way to new technologies so steeped in the mysteries of science, and of such a scale as to outstrip the budgets of many nation-states, that both could only enhance corporate (both public and private) as against individual power.

These essays first appeared in a volume published abroad by the Voice of America and were based on a series of overseas broadcasts designed to celebrate the nation's bicentennial. Richard J. Gordon, at that time forum editor of the Voice of America, originated the idea. Carol Morarsky and Merrill Miller were also extremely helpful in organizing the project and bringing it to fruition.

Carroll Pursell
University of California at Santa Barbara
February 1981

1 Technology in America: An Introduction

Carroll Pursell

The year 1776 saw two great revolutions unfolding. In Europe, the Industrial Revolution was beginning to change both the face of the earth and the lives of people everywhere. In America, a political revolution was bringing forth a new continental empire. The revolutionary generation of Americans was born in a country that labored with a medieval technology in a colonial economy. By the time its work was done, it had deliberately borrowed the new European industrial technology and rejected the Old World's political control. Inevitably those two revolutions became intertwined and both were changed. American innovations produced mass production and enabled mass consumption. Industrial technology directed American energies toward economic growth and continental expansion. Only half consciously, the United States followed the advice of the "Father of Scientific Management," Frederick Winslow Taylor, to "take their eyes off of the division of the surplus as the all-important matter, and together turn their attention toward increasing the size of the surplus until this surplus becomes so large that it is unnecessary to quarrel over how it shall be divided." Only recently have most Americans been led to confront the idea of limits to growth—and the limitations of the technologies we have chosen.

The first Europeans to settle in America brought with them that technology that had been one of the triumphs of the Middle Ages. The square-rigged sailing ship (its course set by the compass and sides bristling with cannon), heavy plow, and stirruped horse saddle all made it possible for the newcomers to hold the native peoples at bay and wrest a living from the land. Mills were built to grind grain and saw lumber as well as full cloth and smelt iron, but most crafts were still carried on by hand, and transportation was limited largely to horseback and boat.

Then, beginning with James Watt's improved steam engine of 1763, an astonishing series of new machines and processes began the Industrial Revolution in Great Britain that made much of the rest of the world's work obsolete. In quick order, transportation (canals and turnpikes), the textile industry (the spinning jenny and power loom), and iron making (pit coal, the puddling furnace, and rolling mill) all combined with the rise of the new profession of civil engineer to create an industrial technology about which Americans knew little. First the shackles of British legislation designed to impede American industry, and then the coming of war itself and lack of transatlantic intercourse, kept the developing nation in ignorance.

With the winning of independence and reestablishment of normal relations with Europe, the founding generation discovered that (as it seemed to it) a wise providence had provided it with a new technology capable of supporting its new nation. With abundant resources but few people to exploit them, Americans who aspired to surpass quickly the splendor and power of the old British Empire soon realized that machines would have to replace hands if the job were to be done. A patent law was passed in 1790, but more important, there was almost an explosion of what came to be called "Yankee ingenuity."

Typical of this startling burst of inventive activity was the automatic flour mill of Oliver Evans. In 1785, he put in operation a mill that moved grain and flour about by means of buckets on belts and endless screws. So fundamental was this innovation that it remained standard in American mills for nearly a century—until the introduction of the Hungarian (roller) process in 1870. At about the same time, two inventors, John Fitch and James Rumsey, built and operated steamboats on the rivers of the East Coast. By the time of Robert Fulton's triumphant *Clermont* voyage from New York City up the Hudson River in 1807, over a dozen American inventors had built experimental steamboats.

One of the nation's most basic contributions to modern technology, the so-called American system of manufacture of interchangeable parts, was also one of its earliest. Despite precursors in Sweden, France, and England, it was in the U.S. national armories (and reputedly in the private factory of Eli Whitney) that the use of machine tools to fabricate small arms from interchangeable parts was first placed on a practical and continuing basis. Within a few years, the practice had spread to the manufacture of clocks, door locks, sewing machines, and similar devices made up of a large number of small metal parts. The dramatic lowering of costs through the American system made all these devices available to a class of citizen that could not otherwise hope to enjoy them.

The sewing machine, patented by Elias Howe in 1846, was a key device on several counts: not only did it become one of the major technologies effecting the role and status of women, it greatly stimulated the ready-to-wear garment trade. In addition, it served to pass the technique of mass production on to such later important industries as bicycle manufacture, and through it in turn to the automobile industry. By 1860, some 110,000 sewing machines a year were being manufactured and sold.

Two other American devices of these years deserve particular mention because they reflected a primary need of the nation—dealing, somehow, with the great scale of the country. In 1834, Cyrus Hall McCormick patented his reaper, and at the first world's fair, the London Crystal Palace Exhibition of 1851, it was said of this device that it was "the most valuable contribution from abroad, to the stock of our previous knowledge, that we have yet discovered." By applying horsepower to the difficult task of cutting wheat, McCormick made possible the rapid growth of the great wheat-producing regions of California and America's "Middle Border."

The vast distances that the United States presented dictated the direction of much of her inventive effort. McCormick's reaper was designed for large farms with few workers. The electric telegraph of Samuel F.B. Morse (first successfully operated in 1844) was another device designed to break down the isolation of the distant parts of the nation. It was, in addition, the first large-scale and commercially important use of electricity.

Along with the telegraph, the railroad was a major factor in linking together the American continent. Although an English invention, the steam railroad was quickly seized on by Americans, improved in numerous ways, and used to tie together a national market for commerce. The completion of the first transcontinental line in 1869 was a triumph of both engineering skill and organizational management.

The needs of commerce and politics also encouraged the invention of a set of machines designed, in one way or another, to facilitate communications. The typewriter (1873) mechanized the task of correspondence and created a job category (secretary), which soon was opened to and then dominated by women—quickening their employment outside the home in growing numbers. The typesetter was closely allied to the typewriter, and together with the steam-powered, "lightning" rotary presses of the same period, led to a golden age of inexpensive, popular newspapers. In 1876, Alexander Graham Bell, a researcher into the problems of the deaf, developed his telephone. It, too, was a business machine at first.

Important as these inventions were in building up the nation, and cementing its economic and political unity, they were quite traditional in the sense that they depended in some degree on the flash of genius and heroic entrepreneur. The nineteenth century, however, was rapidly becoming one in which scientists were able to demonstrate a growing measure of control over natural forces. With new scientific competence, and confidence, came a set of institutions designed to stimulate, transmit, and apply this new technology. The U.S. Military Academy at West Point (1802) was the nation's first engineering school, but it was joined later by many private polytechnic institutes and public engineering schools. The first engineering society, the American Society of Civil Engineers, was established in 1852, and followed by the American Institute of Mining Engineers (1871), American Society of Mechanical Engineers (1880), American Institute of Electrical Engineers (1884), and others even more specialized. In these schools and societies, the beginnings of scientific technology grew out of a living blend of experience and experiment. National legislation in 1863, during the American Civil War, led to the establishment of an agricultural and mechanical school in each state of the Union, and in that same year the National Academy of Sciences, was chartered in the hope of providing the federal government with the best-available scientific and technical expertise.

The new and characteristic technology of the twentieth century came more from organized scientific research than had previous innovations, in part thanks to these new professional institutions. Thomas A. Edison had used a well-equipped and staffed laboratory to produce such late nineteenth-century marvels as the incandescent electric lamp (1879), but the building of a corporate industrial research laboratory in 1901 by the General Electric Company (successor to Edison's own firm) marked the real beginnings of industrial research in the country. Organized by Willis R. Whitney, the laboratory was able to make rapid improvement in the efficiency of electric lamps and employed well over three hundred people by 1918. So good was the science carried on in the laboratory that in 1932, Irving Langmuir was awarded the Nobel Prize for work done there in surface chemistry.

By 1917, the number of industrial research laboratories in the country had risen to 375, and in 1931 there were over 1,600. Private industrial efforts were supported by large scientific attempts on the part of many agencies of the federal government—the Bureau of Mines (1910), National Bureau of Standards (1901), and National Advisory Committee on Aeronautics (1915), to mention only a few. In most states, agricultural and engineering experiment stations (financed in part by the individual states and in part by the federal government) carried on economically important research, while universities attempted at the same time to create new knowledge, pass it on to students, and help society to apply it in a practical manner.

This institutionalization of innovation was matched by attempts to rationalize the other factors of production of American industry. Scientists and engineers, under the leadership of the forester Gifford Pinchot, instituted a conservation movement early in the century to put resource exploitation on a managed and efficient basis. This was particularly critical in terms of energy fuels. Since the birth of the American Republic, reliance has shifted from renewable (and relatively nonpolluting) energy sources such as wind, water, wood, and animals to nonrenewable sources such as coal, oil, and natural gas. At the same time, total energy consumption increased approximately fivefold from 1850 to 1900, and was to be eight times larger in 1975 than it had been in 1900. An increasing amount of this energy, of course, was consumed in the form of electricity. Networks of electric power supply spread rapidly, especially after the federal government established the Rural Electrification Administration in 1935.

During the same period, sparked by Taylor's classic study *The Principles of Scientific Management* (1911), the concept of efficiency was applied to human-machine interactions in the factory environment. Just as the flash of inventive genius was replaced by planned scientific research, and the natural forests and free-flowing streams were replaced by managed tree farms and waterways controlled by engineering, so now were traditional craft skills replaced by scientifically defined and determined tasks.

The productive culmination of these various attempts at rationalization and scientific management was the assembly line, introduced into his automobile plant by Henry Ford in 1914. Based on the principles of the division of labor, delivery of work to the workers, and "planned orderly and continuous progression of the commodity through the shop," Fordism (as it came to be called) became not only the model for twentieth-century industrial production but the symbol, through Charlie Chaplin's film *Modern Times* (1936), of the dehumanization of industrial technology as well.

The entrance of the United States into World War II in 1941 led inevitably to a new federal concern with this, by now, large and productive scientific-technological capability. The Office of Scientific Research and Development, established in 1940 and headed by Vannevar Bush, directed a massive effort to improve the implements of war and, as one historian has pointed out, succeeded in creating a totally new "electronic environment" for war. Such devices as radar (developed in Great Britain) and controlled nuclear reactions presented the postwar world with new technological potentials. The computer, systems theory, jet propulsion, and DDT were only a few of the marvels pushed forward by war. At the same time, the government moved to expand and enhance the capability of universities in basic science. The plan to place an astronaut on the Moon, announced in 1961 and accomplished in 1969, was a symbolic affirmation not only of the new technologies but also a continued confidence in people's abilities to shape them to their purposes.

As the United States enters its third century, it is quite properly turning its attention to the costs as well as benefits of its legacy of scientific and technological prowess. Recent awareness of the limits of growth, both in terms of reduced resources and increased pollution, has forced a closer look at both old and new technologies. The establishment of an Office of Technology Assessment in 1972 was only one indication of the realization that machines and processes are not always and in all ways beneficial. As the nation looks back to assess its past and forward to chart its future, it is certain that science and technology will play a crucial role in whatever future it chooses to pursue. They will be a science and technology more carefully tuned to human needs and limitations than previously, but confident still of their ability to serve the national purpose.

2   The Artisan during America's Wooden Age

Brooke Hindle

Early American technology depended primarily on the skills of its artisans. Most of these skills were ages old, as were the basic tools of the trades. Craftsmen who crossed the Atlantic transmitted these skills to America in the first instance, and an apprenticeship system that taught skills by example then perpetuated them. This process was supplemented by a continuing influx of European artisans as well as by books, manuals, and occasionally visits of Americans to Europe.

Skill was the great requirement in this pattern of technology. For example, anyone who had seen a barrel knew how the staves had to be curved and shaved to produce a tight vessel. The design was wholly without mystery. Yet the skill of eye and hand required could not be quickly approximated by the uninitiated. The slightest error in the angle given the edge meant a leaky container. Similarly, the objective of the wheelwright in making the well-known wagon wheel was apparent to everyone. The fabrication of a satisfactory wheel, however, required practice and skill in making the parts, and then putting them together.

High levels of skill were attained and practiced in a large variety of crafts carried on in colonial towns and cities. On the other hand, most of the early population was rural, and farming was the predominant occupation. Farmers were often separated by large distances from the skills of the cooper, joiner, and cabinetmaker—to say nothing of the silversmith or gunsmith. In fact, even when farmers lived close enough to a town boasting such crafts, the price of the product was more than most of them could afford. As a result, many in America were handy at a variety of handicrafts but skilled in few. As best they could, they built their own barns and houses, made their own furniture and clothing, fabricated many of their own tools and utensils, and bought from practiced artisans only what they could not improvise.

In this setting, American technology did not differ strikingly from that of Europe, but in one respect, Americans enjoyed a great advantage. Especially by comparison with the mother country, Great Britain, Americans confronted a wonderful plenty of wood. The first colonists did not—as is sometimes imagined—find an entire continent covered by a climax forest. Even along the seaboard, the forest cover was broken at many points. Nevertheless, fine trees of all sorts abounded, and throughout the early period those who pushed forward the frontier continued to encounter new forests. By the end of the colonial period, the price of wood had risen in the eastern cities, but for most Americans, wood was plentiful in a measure unknown to contemporary Britons.

The ready availability of wood gave Americans advantages that have seldom been clearly appreciated. Wood was a basic foundation of the economy. Most obviously, it constituted the primary building and manufacturing material. Wooden houses and buildings of all sorts were used to a degree unknown in Britain or the cities of Europe. Furniture, furnishings, boats, ships, wagons, carriages, and all manner of implements were made of wood. Second, wood was the primary fuel of this era. It was used for space heating and, albeit inefficiently, cooking; it was used, sometimes in the form of charcoal, as the source of the heat required in all metallurgical and industrial processes. Third, it was an important chemical product, such as in the form of potash, an industrial alkali; charcoal for gunpowder; and oak bark for the tannin needed in tanning leather. No other substance was as crucial in the technology of the time as wood.

The American supply of wood conferred many advantages, a few of which had certain negative aspects as well. Iron, strikingly, was one of the most important products of the wooden age. It was produced by heating iron ore with charcoal and limestone flux in a furnace. Because England had been so denuded of her trees, she came to a point where she was unable to exploit her still-rich iron mines but had to depend heavily on northern Europe for much-needed iron. Her American colonies had both iron ore and wood in plenty, with the result that iron production was encouraged and became successful. Yet when England developed coke smelting, the colonies did not follow suit because they still had plenty of wood, and charcoal iron was a better product than coke iron. Unfortunately for the Americans, coke smelting in Britain soon led to other technological improvements and was integrally linked with the emergence of the Industrial Revolution. Early in the nineteenth century, Americans lagged in their industrial development precisely because their plentiful wood led them to cling, anachronistically, to charcoal iron.

In many applications of wood, the American advantage of plentiful supply seemed unalloyed. Shipbuilding, for example, was one of the most essential enterprises of the time. Wood was the basic raw material required, and England was much disadvantaged because she did not have the wood required to build the ships needed to maintain her extensive commerce. She came to depend on America, which supplied not only masts and timber but American-built ships too. America also built her own ships—for commerce, whaling, and fishing. By the time of the American Revolution, one-third of the ships flying the British flag were American built. American commerce and prosperity rested heavily on the ships she had learned to build so well because she had so much available wood. But for her shipping and iron, which she then produced in greater quantity than the mother country, America would never have been able to win her War of Independence.

The technology of shipbuilding was typical in the manner in which it rested on individual skills. It depended on a complex of arts or crafts that were learned by experience in an apprenticeship pattern. The development of extensive shipbuilding capabilities in the larger cities encouraged the growth of the spectrum of crafts that were required in completing a ship. Indeed, a large proportion of the crafts available in a seaport town, such as Newport, Rhode Island, had to be called on to complete an oceangoing ship, the largest technological enterprise attempted in the early period. Not only shipwrights, carpenters, and caulkers were required but also joiners, blacksmiths, tinsmiths, instrument makers, painters, glazers, ropewalks, and sailmakers. Many artisans who did not even make or work with wooden products were sustained by the great shipbuilding effort that rested squarely on the supply of wood.

In the colonial era, when the craft skills on which this wooden age technology rested were being assimilated, American inventiveness was not notable. Changes occurred only in tools, techniques, and designs slowly, and many of the differences noted between European and American patterns were explainable by differing needs or differential supplies of materials or

The Flying Cloud, one of the fastest of the wooden clipper ships built in nineteenth-century New England shipyards, could cover 433 statute miles in one day.
Source: National Archives

skills. In America, for instance, many more structures were built of wood than of stone or brick as in Europe. In New England especially, where building stone was lacking and good brick clay was scarce, wood construction was nearly universal except in the larger cities. Wooden structures served along most of the frontier. Tile roofs, introduced at points, quickly gave way almost everywhere, predominantly to wooden shingles or thatch but sometimes to slate.

Such changes and improvements as did emerge were sometimes silent or evolutionary, as in the case of the elegant American ax. The emergence of the American ax is undocumented and cannot be attributed to any known person or persons. It was the product of small changes as well as, no doubt, much trial and error. It represented a remarkable improvement. The American ax was more compact than its European prototype; it was beautifully balanced so that the poll and bit were nearly equal in mass, and the center of gravity was near the centerline of the handle. The handle was not a straight pole any longer but instead was gracefully curved—and indeed often fitted to the height and swing of the individual axman for whom it was made.

The result was a dramatic improvement in the speed with which one could fell trees. Trials demonstrated that an American with his ax could fell three times as many trees in a given time as someone with a European ax. The improvement cannot be attributed to higher levels of inventiveness in America but rather only to the great opportunity offered in the vast increase in the use of the ax in the New World.

A similar but much broader-based technological achievement was recorded in the nineteenth century when the United States seized leadership in the development of woodworking machinery. Powered woodworking machinery had been a part of the American scene from an early date. Sawmills were second only to gristmills in number during the colonial period; they depended primarily on waterpower, and provided large quantities of sawed lumber not only for the colonies but also for export, especially to the West Indies. Again, the sheer quantity of lumber cut and used was the ground on which many American innovations were based.

Gang saws, muley saws, and other refinements on the sawmill followed. Shapers, planers, and devices for finishing wood were developed further and faster in America than elsewhere. Some of these were specific products of individual inventiveness, as for example, the Blanchard lathe. Invented by Thomas Blanchard, it could make automatically a complex wooden form—following the pattern of a model. In this case, gunstocks could be produced much more quickly and with less skill on the part of the operator. The principle was quickly and widely extended.

Another specific invention that grew out of the vast plenty of wood in America was the balloon frame house. Augustine D. Taylor introduced it in 1833 in Chicago. Instead of the heavy timbers that had to be carefully fitted and joined in preceding house framing, the balloon frame was made by using many lengths of light lumber—ultimately by two-by-four-inch boards rather than by ten-inch-square timbers. This effected great economies in the amount of wood used. It also speeded the job of framing a house and, again, reduced the skill required. A balloon frame house was made by simply nailing the pieces in—eliminating the laborious process of cutting mortise and tenon joints, or drilling and fitting trenails.

This procedure of mechanizing processes and simplifying the work remaining to be done by the artisan was directly encouraged in lumbering

Wooden balloon frame housing is quickly built and requires relatively simple skills as demonstrated by these Amish farmers working on a home in Indiana.

and woodworking. American advances in this direction were notable because of the plenty of wood along with the large part played in the economy by wood and wood products.

At the same time, other forces encouraged mechanization and simplification. Americans received rapidly the advances of the Industrial Revolution, which coincided in some of its key beginnings with the era of the American Revolution. They were especially receptive to the mechanization of textile manufacturing and introduction of the steam engine. Indeed, they were developing their own significant inventions and innovations— in steam engine construction by the first decade of the nineteenth century, and in textile machinery by the 1820s. Then, on their own, the Americans pioneered in the development of interchangeable parts in combination with the factory system to open the access to mass production in what the British called the American system of manufacturing.

The great advances in woodworking machinery were achieved by applying the developing techniques of mechanization to this great area of need— woodcutting and shaping. The plenty of wood set the stage for one of the most striking applications of the American system: clock making. The use of interchangeable parts had arisen in gun manufacturing, and one of its primary areas of development was in the Connecticut River Valley. Here, Eli Terry made a similar move from handmade brass clockworks to wooden works in which the wheels were cut to standard patterns in large quantities. These wooden, mass-produced clocks, because they were so cheap, rapidly became dominant in the market. The next step, taken most successfully by Chauncey Jerome, was to make brass clockworks on the basis of inter-changeable parts—a more difficult process that would have been hindered if the transition through the easier construction of mass-produced wooden works had not occurred.

Blanchard, Taylor, Terry, and others were artisans or craftsmen who proceeded in much the same manner as artisans had for centuries. They learned a trade that they practiced successfully, beginning with the tools and techniques commonly used in their time. In their initial work, they copied familiar models—the age-old method and objective.

What led them to break out of established patterns of production to invent new, simplified, mechanized ways for making products—and change the products in the process? The opportunities in this period of history, and especially in America in wood-based technologies, encouraged them to invention and innovation. The strongest line of encouragement led to simplification of process and product as well as mechanization. This was, indeed, the same sort of encouragement that had led a little earlier in Britain to the classical Industrial Revolution along with achievements that had great appeal and influence in America.

Yet the encouragement to invention and innovation did not always lead to mechanization. In the wood-based technologies, a more ancient sort of change is exemplified in the construction of truss bridges, at first made entirely of wood. Such bridges were used in road construction for crossing streams, but American road construction long remained underdeveloped and inferior. The big need for cheap truss bridges—and strong monetary encouragement to invention and improvement—arose with the great rail-road building effort. American railroads, covering vast stretches of geogra-phy, had to be built as cheaply as possible. Tight curves and steep slopes were accepted, and cheap bridges were preferred, to expensive cutting and filling or route detours.

The result was to encourage inventors to invent new truss forms that would be strong enough and cheaper than those being built. The objec-tive was to use less material and, second, less labor. For a time, a great number of patents were issued for new trusses, with each inventor hoping that his invention would be widely adopted and make him rich. New and better trusses were devised in a process of simplification and economy of construction—but not of mechanization.

In another wood-based technology, shipbuilding, American encouragement to change and improvement also failed to produce mechanization, but as in the case of truss bridges, spurred what should be called design improvement. Americans had much experience in the construction of wooden ships and wooden boats for inland transportation. In each case, specifically American developments were often keyed to specific needs. For example, the Durham boat, a large, low-draft craft for carrying bulk cargoes, was used on the eastern rivers. The western keelboat had similar missions. The Baltimore clipper was a specifically American sailing craft, well designed for the coastal sailing intended, and in the later international clipper ship competition, the Americans registered significant design achievements.

Perhaps the kind of change encouraged in shipbuilding is best epitomized in the design of the US frigates begun at the end of the eighteenth century. The overall concept of these ships was that of the leading American shipbuilder, Joshua Humphreys, who produced something different from either the French or British prototypes. His forty-four-gun frigates were larger and more heavily armed than similarly rated ships, and they had fine lines, which made them better sailors. They were creative innovations in the classical sense of representing improvements over existing models—rather than differences in kind.

Artisans, then, came out with many kinds of change, invention, and improvement based on America's wonderful plenty of wood in combination with the variety of needs and opportunities that arose. They produced a great number of striking design improvements keyed to local circumstances. Many of their innovations did contribute to the nineteenth-century movement toward simplification and mechanization. Invention was heavily encouraged, and American artisans were often led to think more about change and improved production than about excellence of finish or craftsmanship.

Certainly this process fed many American characteristics. It has been suggested, for example, that the "replacement economy"—that is, the manufacture of cheap goods that could be replaced after a short period of use instead of being repaired or used in a worn state—may have grown out of the experience with wood-based technologies. Also, because wood was plentiful, the Americans did not think much about waste. They used saws that ran faster and cut more efficiently than any predecessors—unconcerned that they wasted more in sawdust than if they had used other saws.

The experiences with wood changed America, the nature of many technologies, and the world. They changed too the craftsmen who were themselves the agents of these changes. Today more than before, we are aware that the improvements were purchased at the cost of some real disadvantages. All changes were not for the better ultimately. In their day, however, they represented great achievements in which the negatives were relatively unimportant. They played a major role in leading us to our present world.

3 Thomas Jefferson and a Democratic Technology

Hugo A. Meier

The election of Thomas Jefferson as the third president of the new United States of America in the so-called Revolution of 1800 marked the fulfillment of the republican ideal in the Western world. Embodied in that achievement was a tacit recognition of the American faith in people's capacity to wed a political ideology of freedom for all people with the practical goal of uniting advances in knowledge to the well-being of the ordinary citizen.

Jefferson, republican ideologue, sophisticated citizen of the late eighteenth-century world of science and letters, remains for us, in our technology-dominated twentieth century, America's earliest distinguished spokesman for a democratic technology.

Jefferson's own Virginia background did not seem propitious for what we in the twentieth century conceive to be the democratic society. Here existed a close-knit, elitist culture rooted in the conservative traditions of an agrarian-based social order, itself strongly dependent on human slavery. But America in 1776 was still too new a land to have fostered any genuine aristocracy. Indeed, Peter Jefferson, Thomas's father, had been obliged to make his own way as surveyor, and succeeded—as so many of the Virginia gentry were to do—by the shrewd acquisition of baronial acres bargained out of the vast reserves of forest and farm lands that the New World afforded.

America was then already the land of opportunity for those with initiative. Judicious marriages with established families and their generous landholdings provided the obligations, incentives, and means by which an excellent education and experience in public affairs might be acquired and a responsible sense of paternalism developed. Thomas Jefferson shared this patrimony of landed wealth and respect for knowledge. He, as had George Washington, acquired the surveyor's skills. By inheritance and assiduous accretion of acres, Jefferson's lovely estate on the Rivanna River, Monticello, reached an impressive size. Its master joined not only the eighteenth-century world of refinement and culture but also participated avidly in its intellectual pursuits.

For Jefferson the scholar and political ideologue, Europe and in particular France constituted a congenial spiritual home. Appointed as Benjamin Franklin's successor as minister to the court of Louis XVI, Jefferson established an easy rapport with French men of science and letters. He was less happy with England and the English culture despite his own intellectual debts to both. The burgeoning Industrial Revolution in England, however, left Jefferson with considerable admiration of British technology as well as disenchantment with its less happy urban and human consequences. He was keenly aware that the focus on experimental science, which fascinated French philosophers, in England was directed at converting labor and resources into useful goods for everyone.

At home, the colonies sought and found their freedom in the midst of embarrassingly rich natural resources. These subtly shaped popular tastes and entrepreneurial talents toward satisfying domestic wants. Science per se was left to the preoccupation of leisured clergymen and other amateurs, or confined to the lecture rooms of America's embryo colleges. Technology was for most colonials the natural response to the practical demands of their environment. The "know-how" approach to knowledge, rather than the "know-what" one, fell quite naturally within the interests of the pragmatic Americans. They could borrow freely from the knowledge and expertise of Europe—a function that Jefferson also served via his tours abroad, his beloved books, and the streams of knowledgeable visitors who climbed the slopes to Monticello.

As a member and president of the American Philosophical Society, Jefferson acted as intermediary between its members and the foremost European men of science. His vast correspondence addressed to knowledgeable men on both sides of the Atlantic reflected in its diversity and devotion to learning the spirit of the volumes of Denis Diderot's illustrious *Encyclopédie*, copies of which Jefferson sought out for grateful American colleagues. His own library, despite losses by fire, represented a cross section of Enlightenment knowledge. When British armies burned the national Capitol in Washington in 1814 and its library of perhaps three thousand volumes, Jefferson contributed eleven wagonloads of volumes from his own library—about sixty-five hundred books—as the nucleus of a new Library of Congress.

Such devotion to learning was not uncommon even amid the war-harried and empirically minded scholars of the New World. That a people traditionally practical and devoutly religious might have reservations about the strange social and political dogmas of the French Enlightenment was also apparent. They had lost enthusiasm for continuing revolutions in a renewed quest for social stability and economic success, and Jefferson's affinity for French revolutionary philosophies and religious skepticism encouraged the beliefs of his Federalist Party opponents of his singular unfitness for public office.

As a man of the Enlightenment, Jefferson was not content with accepting the world unquestioned. The universe itself now was a solvable problem; Isaac Newton had seen to that. If Jefferson was intensely curious about that universe and its laws, celestial mechanics, and magnificent global Earth, that curiosity was aimed at understanding its mysteries in ways that might lead to the improvement of the human condition. "Perfect happiness, I believe, was never intended by the Deity to be the lot of one of his creatures in this world," he observed in 1763, "but that he has very much put in our power the nearness of our approach to it, is what I have steadfastly believed."

Thomas Jefferson (painting by Rembrandt Peale).
Source: New York Historical Society

The eighteenth century with its faith in science and reason was appropriately the century of republicanism as a favored political expression of both concepts. "Science," said Jefferson, "is more important in a republican than in any other government." Not only was a republican environment especially congenial to science and technology, he believed, but the combination also contributed to the cultivating of a new sense of American nationalism. Granted, as Jefferson had said, that "in an infant country like ours we must depend for improvement on the science of other countries, longer established, possessing better means, and more advanced than we are," the New World offered unparalleled opportunity for scientific and technological discovery. "It is the work to which the young men who you are forming, should lay their hands," he advised President Joseph Willard of Harvard College. "We have spent the prime of our lives in procuring them the precious blessing of liberty. Let them spend theirs in showing that it is the great parent of *science* and of virtue; and that a nation will be great in both, always in proportion as it is free."

The new national pride of the versatile scholar of Monticello kept him wary of foreign claims to inventions that he believed to be of American origin—and especially so where the patronizing mother country was involved. The new Patent Office was for Jefferson a symbol of American creativity, and he confidently advised William L. Thornton, its administrator, "I think the English must acknowledge we excell them in the faculty of invention. There can be no better proofs than are furnished by our office." He shared a common American suspicion of the employment of European engineers and technicians by the new government because the practice would inhibit the development of native American talents to the detriment of a truly national technology.

Jefferson's sense of American nationalism sometimes goaded him into exaggerated rebuttals of European criticisms of the poverty of American achievement in science and invention as well as unacknowledged borrowings from abroad. He pointed out that by comparison of populations, France might be expected to show six geniuses and England at least three for every American genius. But he also believed in the internationalism of science, refused to restrict in any way the foreign use of his own inventions, and urged that nations reorganize their international relations with the same harmoniousness characteristic of scientific societies, which constituted "a great fraternity spreading over the whole earth."

If Jefferson believed science and technology to be important to the success of a republican form of government, he was even more concerned with their impact on the everyday life of the citizens of that republic. Although the term democracy was not on the popular tongue of Americans during most of Jefferson's life, his faith in the common man found expression in his concern with applied science and social change. In a democracy, flexibility and change were conditions most favorable to technological advancement. Reason, so well exemplified in scientific progress, must replace superstition. Knowledge must replace ignorance. Conservatism must accommodate to receptiveness toward social change. He accused the New England clergy of insisting on looking back too exclusively to the opinions and practices of their forefathers "instead of looking forward for improvement." To Robert Fulton, recently acclaimed for his successful steamboat, Jefferson wrote in 1810, "I am not afraid of new inventions or improvements, nor bigoted to the practices of our forefathers. It is that bigotry which keeps the Indians in a state of barbarism in the midst of the arts."

If reason and talent were to be at the service of every man, then utility should be the final judge of merit in a democratic technology. This capacity to serve the needs of the average citizen also served the highest humanitarian goals. The line between pure science and technology was blurred in the new republic; a qualified scientist such as the mathematician-astronomer David Rittenhouse felt sometimes burdened, but seldom offended by his assignment to duties of clock making or surveying. Jefferson planned to introduce branches of technologically oriented education into the curriculum of his proposed University of Virginia, observing that "I have taken some pains to ascertain those branches which men of sense, as well as of science, deem worthy of cultivation." Existing schools, he complained, taught their versions of science "too much *in extenso*, for the limited wants of the artificer or practical man."

In his concern for applying science to useful ends, Jefferson invites inevitable comparison with Franklin. Both men shared the Enlightenment curiosity about man and his world. Both favored a practical view of science. Jefferson shared the point of view that Franklin expressed in advice to a correspondent in 1783 who was "endowed with a genius for the study of nature." Franklin wrote, "I would recommend it to you to employ your time rather in making experiments, than in making hypotheses and forming imaginary systems, which we are all apt to please ourselves with, till some experiment comes and unluckily destroys them." Franklin drew from *Poor Richard* his example of the ingenious striking sundial, which by action of the sun on twelve burning glasses and trains of gunpowder, would fire off seventy-eight guns in proper sequence: "Let all such learn that many a private and many a public Project, are like this Striking Dial, great Cost for little Profit."

Although admitting that he was himself an empiric in natural philosophy—"suffering my faith to go no further than my facts"—Jefferson did acknowledge the usefulness of hypothetical speculations "because by the collisions of different hypotheses, truth may be elicited and science advanced in the end." Franklin proved to be the better scientist—a close observer and experimenter in natural phenomena as well as the practically minded inventor of improved fireplaces. Jefferson hoped to use his retirement for the sustained study of natural science, but in his advanced years he recognized that science had progressed considerably beyond his capacity to catch up with it.

His pragmatic concern for using science to enhance human comforts made Jefferson a close and curious observer of things useful. Friends and acquaintances drew freely on his generosity while he was abroad to obtain for them those things helpful to the citizens of his republic. His self-professed "zeal to promote the general good of mankind by an interchange of useful things" resulted in a stream of ideas, books, commodities, and inventions likely to improve American technological know-how. What might appear to be mere gadgets to some—improved letterpresses and efficient cylinder lamps burning olive oil—were, for Jefferson, instruments serving the inquiring mind. He found matches to be "a beautiful discovery, and very useful" in such humble tasks as lighting candles, kindling fires, and sealing letters.

Jefferson staunchly defended the dedication of ingenuity supported by a "practical" science to domestic purposes. It represented for him the democratic goal of a useful science, however humble those uses. "You know the just esteem which attached itself to Dr. Franklin's science, because he always endeavored to direct it to something useful in private life," he advised Dr. Thomas Cooper at Carlisle (now Dickinson) College when he planned a course of instruction in chemistry in 1812. "The chemists have not been attentive enough to this. I wished to see their science applied to domestic objects, to smelting, for instance, brewing, making cider, to fermentation and distillation generally, to the making of bread, butter, cheese, soap, to the incubation of eggs, etc. And I am happy to see some of these titles in the syllabus of your lecture. I hope you will make the chemistry of these subjects intelligible to our good housewives."

He approved the new steam engine of one George Fleming in 1815 because it was simple, cheap, and adaptable to the most common purposes. "A smaller agent, applicable to our daily concerns, is infinitely more valuable than the greatest which can be used only for great objects. For these interest the few alone, the former the many." He advised him to develop "a domestic machine" to be used for lifting water, washing linens, kneading bread, heating the hominy, churning butter, and turning the spit. If these seemed trivial, he argued, "of how much more value would this be to ordinary life than Watt's and Bolton's [sic] thirty pair of millstones to be turned by one engine." A little self-conscious about his improvements to the common plow, Jefferson felt obliged to apologize to Sir John Sinclair, president of the National Board of Agriculture in London in March 1798, that "the combination of a theory which may satisfy the learned, with a practice intelligible to the most unlettered labourer, will be acceptable to the two most useful classes in society."

But behind utility lay Jefferson's wide-ranging curiosity and richly eclectic knowledge. Monticello's halls and walls, its very architectural design, its kitchens, workshops, and household furnishings, all reflected the diversity of his interests. His library was rumored to contain multivolumed studies of prairie dog genealogy and the agility of the flea. He was as much interested in the physical and acoustic means by which his fiddle produced music as in the music itself. A consuming interest in mathematics persuaded him to decimalize the new republic's currency, and reform its inherited archaic system of weights and measures, "bringing the calculations of the principal affairs of life within the arithmetic of every man who can multiply and divide plain numbers."

As citizen of a militarily weak nation hovering behind an extensive seacoast and enjoying fine harbors, Jefferson shared a prevailing defensive stance in foreign policy. Thus he welcomed Fulton's invention of "torpedoes" or underwater mines, observing that "I sincerely wish the torpedo may go the whole length you expect of putting down navies." Jefferson, however, also supported the idea of an efficient American navy capable of defending harbors and consisting mostly of inexpensive gunboats. At least 176 such vessels were constructed during his presidency—each ranging from fifty to seventy feet in length, visually unimpressive in light of the splendid vessels of His Majesty's navies, and nicknamed by his critics as "Jefferson's democratic sinking fund." He established a corps of engineers in 1802 as the nucleus of the West Point Military Academy, favoring technicians over military men. And Jefferson shared the contemporary fascination with balloons, which he believed, as did Franklin, might make war too hazardous for aggressors by exposing their own territories to aeronautical invasion.

Jefferson's ready ear for reports of novel discoveries and unique inventions led to the apocryphal story that he himself had invented an elevated sawmill operated by the power of wind on sails, and had one constructed with theoretically good judgment on a hilltop, but had failed to consider the more practical problem of getting the heavy saw logs up to it. Yet the master of Monticello was concerned with more than visionary sawmills. As vice president, he dispatched two emissaries to Europe for the purpose of reporting "Objects of Attention for Americans" in the field of European technology. Everything of interest to agriculture must be closely observed as well as developments in the "mechanical arts so far as they reflect things necessary in America, and inconvenient to be transported thither readymade, such as forges, stoves, quarries, boats, bridges." But as regarded the lighter mechanical arts and manufactures, some only might be worth a superficial view because "circumstances rendering it impossible that America should become a manufacturing country during the time of any man now living, it would be a waste of attention to examine these minutely." A majority of agrarian-minded Americans shared, to be sure, that judgment. While in France, though, Jefferson was fascinated in 1789 by early experiments with interchangeable parts, and sent samples and instructions to General Henry Knox, the secretary of war, with a vain hope that the inventor and his workmen might remove to the United States. In England, he interviewed Matthew Boulton about the comparative efficiency of his steam engines, and arrived at the impressive judgment that "a peck and a half of coal perform exactly as much as a horse in one day can perform." Meanwhile, he continued to collect and ship to America new books of great variety in which European know-how and ideas could be made more widely available to Americans.

Jefferson foreshadowed the later worldwide reputation of Americans as lovers of mechanical gadgets. A faithful and perhaps obsessive correspondent, he was delighted with the "polygraph," a unique copying instrument invented by a Philadelphian. The polygraph consisted of a desk above which was attached a mechanical movement to which from three to five pens were suspended. Jefferson was exceedingly pleased with his new ability to write several "original" copies of letters while the device faithfully imitated his own pen strokes in making duplicates. His urge for perfection led Jefferson to attempt improvements, observing that "it gave me so much satisfaction that I thought it worth while to bestow some time in contriving one entirely suited to my own convenience." He called it his "portable secretary" and "the finest invention of the present age"—a notion not necessarily shared by Americans of a lesser epistolary enthusiasm, for the device did not become widely popular.

Agriculture was still the overwhelming preoccupation as well as occupation of most Americans, and Jefferson proposed to elevate it to a science so that all Americans could enjoy the benefits of contemporary experimentation in soil improvement, diversified farming, conservation, and plowing and harvesting techniques. While still secretary of state, Jefferson heard of a new threshing machine on a farm near Philadelphia and went to look at it personally. Coming on a description of a British harvesting machine, Jefferson obtained a model and duplicated it at Monticello. Not fully satisfied, he constantly sought its improvement.

Most notable, of course, was Jefferson's own careful designing of a plow that, based on mathematical principles, was intended to minimize earth friction, and thus conserve horsepower and actually plow several inches deeper with the same expenditure of energy. When in France in 1788, he had noticed the clumsiness of moldboards on plows being employed by peasants near Nancy. He computed the most satisfactory shape into which such an implement might be hewed and later used such a plow at Monticello, where he was pleased to observe that ordinary plows tended to wear into a comparable shape—practical evidence of its functional success. Jefferson did not patent the invention but instead was satisfied with public recognition of his achievement, including a gold medal from a French agricultural society and honorary membership on the English Board of Agriculture.

The desire for self-sufficiency at Monticello interested Jefferson in any process or device that would contribute to that goal. Being responsible, as secretary of state, for registering patents, he received Eli Whitney's required drawings of his cotton gin invention in October 1793 and hastened off a

Jefferson's home, Monticello, which he designed and built himself near Char-lottesville, Virginia, beginning in 1769, serves as a showcase for his practical ideas and inventions.

letter to Whitney filled with curiosity about that surprising machine. "One of our great embarrassments is the clearing the cotton of the seed," he explained. "I feel considerable interest in the success of your invention, for family use." He wondered if the machine had been tried thoroughly or was "but a machine of theory"—another hint of Jefferson's concern for utility as the standard of invention.

Household technology offered Jefferson possibly his greatest oppor-tunity to demonstrate his natural ingenuity. Visitors to Monticello have for nearly two hundred years marveled at and often found amusement in the strange, clever, and almost always utilitarian devices sprung from the mind—and frequently contrived by the hands—of the disciple of democ-racy. The list was long and various, and included a basic workshop and its tools. A turnstile-like device provided the convenient hanging and retrieval of coats and other clothing. A swivel chair permitted Monticello's master to obtain a circumambient view of his library-study. Among his adjust-able music stands was one also convertible to a table. A walking stick was expandable into three legs and a cloth seat to provide the convenience of a stool. A special mechanism opened double glass doors simultaneously.

There were other fruits of Jefferson's inventive imagination. The famous Monticello clock presented one face inside and the other face outside the house, used cannonballs as weights, needed winding but once weekly, and not only struck the hours but also indicated the days. A clever weather vane over the east portico, connected by a rod to a dial in the ceiling below, informed the master as well as the neighbors about the direction of the wind. The ceilings at Monticello, too, were fireproofed. Ever the restless wanderer, Jefferson utilized a pedometer to measure his walking tours and added an odometer to his carriage to count off the miles of more distant journeys.

Jefferson's pleasure in acting as host to many and often-distinguished guests inspired an intriguing "serving panel" that expedited the serving of warm food to the dining room from the pantry. A central vertical swivel turned a single-panel door on whose shelves serving dishes might be placed to minimize the intrusion of the traffic and curious ears of servants in the dining area. Dumbwaiters, concealed behind paneling on either side of the mantelpiece, brought Jefferson's cherished imported wines from the cellar to the dining room.

While president and deeply concerned with national defense, Jefferson modeled a dry dock intended to preserve his precious gunboats from storm and rot. Most of the gunboats would be kept in reserve as an early version of the "mothball fleet" idea following World War II: defense coupled with economy. Jefferson proudly kept a model on his mantelpiece despite the tart reception of the concept by his critics.

A major issue confronting the new government was determining the proper kind of support for science and invention. The new Constitution was to provide for protection of invention by a monopoly patent grant for a "limited" number of years. Jefferson had strong but not inflexible notions about such support. Fundamental to his thinking was a belief in the freedom of ideas and dangers inherent in restricting their applicability. In a letter to James Madison from France in 1788, Jefferson earnestly advised the prohibition of even limited monopolies. Madison attempted to convince Jefferson that "monopolies are sacrifices of the many to the few," and since in America power lay with the many, there was little danger that the few would be unduly favored. Jefferson eventually conceded that progress in the sciences and arts would be promoted by securing temporary exclusive rights for original writings and discoveries.

Central to Jefferson's thought was the thesis that ideas were universal and "should freely spread from one to another over the globe, for the moral instruction of man, and improvement of his condition." Nature had intended ideas to be "expandable over all space" like fire, and as "incapable of confinement or exclusive appropriation" as was the air that men breathed. Moreover, once publicly disclosed, a new idea forced itself into the possession of everyone, and the receiver could not dispossess himself of it. "Its peculiar character too," said Jefferson, "is that no one possesses the less because every other possesses the whole of it. He who receives an idea from me receives instruction himself without lessening mine, as he who lights his taper at mine receives light without darkening me." Consequently, Jefferson conceived government's role to be that of disseminating information useful to the citizens rather than ensuring material profit for innovators.

Invention, he also argued, was an accumulative process rather than the single fruit of genius. "The fact is that one new idea leads to another, that to a third, and so on through a course of time until some one with whom no one of those ideas was original, combines all together and produces what is justly called a new invention." Such thinking prompted his annoyance with British claims to what were to Jefferson at least equally American inventions. "I see by the Journal of this morning, that they are robbing us of another of our inventions to give it to the English," he protested in 1787. The "invention" consisted of a method for making a wheel of one piece, which, Jefferson claimed, Philadelphians enjoying Sunday picnics across the Delaware River saw on every farmer's cart. The idea, brought to Europe by Franklin, could be found in Homer—and that, said Jefferson, was where New Jersey farmers probably picked it up "because ours are the only farmers who can read Homer."

Troubled as he was by the demands of inventors under the new patent law of 1790 for grants of monopoly, Jefferson saw as great problems for technological progress under such a system as he did for the likelihood of the encouragement of innovation. Especially irritating were the sweeping claims of Oliver Evans, whose genius in invention was coupled with a strong sense of property rights. Evans's automated gristmill—capable of grinding, sifting, and bagging flour by means of waterpower—was the object of mingled envy and resentment by millers generally. Jefferson had completed his own flour mill in 1806 with such improvements as elevators, hopper boys, and conveyor buckets. Forthwith, he was presented a bill for his use of these Evans-claimed ideas—and paid it, but not without some philosophical grumbling: "If the bringing together under the same roof various useful things before known ... entitles him [Evans] to an exclusive use of all these, either separately or combined, every utensil of life might be taken from us by a patent."

Jefferson also strongly opposed attempts to monopolize the principles underlying classes of inventions and disputed Evans's claims on this basis. "I can conceive how a machine may improve the manufacturing of flour: but not how a *principle* abstracted from any machine can do it." He warned that such practice might eventually result in anyone who applied an ax, hoe, or spade to a new purpose demanding the exclusion by patent of similar use of these essential tools. Indeed, "if a new application of our old machines be a ground of monopoly," he remarked, "the patent law will take from us much more than it will give. ... A man has the right to use his knife to cut his meat, a fork to hold it; may a patentee take from him the right to combine their use on the same subject? Such a law, instead of enlarging our conveniences, as was intended, would most fearfully abridge them, and crowd us by monopolies out of the use of the things we have." The spinning machine, thus, should legitimately be usable for any purpose to which it was adaptable. A Georgetown smith, said Jefferson, had some sixteen years earlier made a device to crush plaster that, in a modified form, was now an Evans monopoly for crushing corncobs.

Jefferson was also unhappy about the practice of permitting sales of patented devices by geographic districts only. "Every man," he said, "should be free to use anywhere what he can lawfully buy anywhere. This abuse with the plagiarisms committed & [*sic*] imposed on us render the advantage of the patent law problematical."

The length of such patent monopolies also concerned Jefferson, who in 1807 wrote Evans that while an inventor deserved the benefits of an invention for a certain time, "it is equally certain it ought not to be perpetual: for to embarrass society with monopolies for every utensil existing, and in all the details of life, would be more injurious to them than had the supposed inventors never existed; because the natural understanding of its members would have suggested the same things or others as good." He did admit, however, that the length of time for a patent might have to be conditioned by the American environment, whose sparse settlement and vast distances naturally delayed the knowledge as well as spread of an invention to the disadvantage of a patentee.

Whatever may have been Jefferson's position philosophically on the touchy issue of patent monopoly, as Washington's secretary of state he was confronted with the practicalities of applying the republic's first patent law. The act, passed in 1790, was to be administered by a board consisting of the secretary of war, attorney general, and secretary of state. Jefferson's other duties, to be sure, already included foreign affairs, the preserving of federal records, and promulgating of the laws, and all to be accomplished by a modest budget and small staff. But recommendations that the board consist of experts were stilled by prevailing fears of a new bureaucracy indifferent toward inventors' rights.

It was Jefferson, therefore, who was principally responsible for applying the new act—not unexpectedly in light of his reputation. His dislike for monopolies was now coupled with a concern for detail and accuracy, and a determination on standards for invention based on their novelty, utility, and social importance. On July 31, 1790, the first patent was issued to one

A copy of the first U.S. patent, issued in 1790, with two of the signers pictured: Jefferson and Washington. This patent was valid for fourteen years, but under a later law, seventeen-year patents were established.

Samuel Hopkins for a new manner of preparing pot and pearl ashes. Only two other patents were issued in 1790. Meanwhile, Jefferson enjoyed his opportunities to meet and converse with men of scientific and technological interests.

The process of issuing patents was tedious. The Board of Arts met monthly to compare notes on applications, requesting proper specifications, drafts, or models. The complex task of determining credit for the new steamboat inventions resulted in four claimants receiving patents. Frivolous and inoperable devices were rejected. "Actual experiment must be required," insisted Jefferson, "before we can cease to doubt whether the inventor is not deceived by some false or imperfect view of the subject."

A classic example of Jefferson's cautiousness and thoroughness as patent administrator related to the application of Jacob Isaacs for a patent on a still and furnace designed to distill freshwater from seawater. In an era of sailing vessels and long, uncertain voyages, such a device was scarcely unimportant. Jefferson required Isaacs to assemble his contrivance in the secretary of state's office for a test. But he also called in Rittenhouse, astronomer-clockmaker-surveyor, and Caspar Wistar, one of America's earliest glass technologists, to help judge the merits of Isaacs's method. After five tests, the latter's request for a patent was denied. Jefferson's own long report pointed out how the idea was not a sufficient improvement over already-known methods of converting seawater.

If his interests seemed universal, Jefferson's time and energy were not limitless. He soon felt oppressed by the mounting duties of his several offices, and his conscience was troubled by his doubtful capacity to judge fairly the increasing variety of ideas and inventions that were paraded before him by the republic's more fertile innovative minds. He reported that he felt "oppressed beyond measure by the circumstances under which he has been obliged to give undue and uninformed opinions on rights often valuable and always deemed so by the authors."

In February 1793, a new patent act greatly simplified the tasks of the secretary of state's office by placing responsibility for judging the validity of issued patents in the district courts. While this move also met the democratic suspicion of government bureaucracies, it additionally resulted in so casual a process of registration of patents that it was claimed individuals copied and repatented ideas from existing models in the patent office. Jefferson, happy to be relieved of the duties of search and examination, hoped that the new approach might lead to a new form of jurisprudence.

During Jefferson's term as administrator, only sixty-seven patents were issued—a sign less of the failure of American ingenuity than it was of the exacting standards of one who suspected monopoly of any kind, and possessed a mind sufficiently cautious and critical to make the patent right a privilege to be achieved only by genuine merit. The basic revision of the American patent system in 1836 resulted in a compromise by including Jefferson's rigorous tactics in a new system of examination to replace the largely clerical system succeeding him.

Jefferson's ideal of a technology dedicated to the popular weal both reflected and enforced an attitude that would increasingly dominate the American culture. The nineteenth century—and Jefferson lived through its first quarter—carried on that ideal in an ever-proliferating abundance of practical inventions for everyday uses.

Alexis de Tocqueville, the French liberal politician and social commentator who visited the United States during the 1830s, found the Jeffersonian ideal thriving in a manner that might have disturbed even Jefferson. A worship of practical invention, concern for utility, and enthusiasm for the material consequences of applied science—all, Tocqueville observed in his still-cogent portrayal of *Democracy in America*, were the democratic fruits of science applied to the needs of the people. "Those who cultivate the sciences amongst a democratic people," he concluded, "are always afraid of losing their way in visionary speculation. They mistrust systems; they adhere closely to facts and the study of facts with their own senses."

The results of such materialistic application were to include a rising standard of living that would remain the envy of the world into our own times. Today while we fear Americans may be nearing the end of a long era of technologically created abundance, nonetheless we join with much of the rest of humankind in lamenting its consequences of environmental despoliation, unimaginable horrors of technically perfected warfare, and the depersonalizing of life itself.

There is, however, still present the historical consciousness that the Jeffersonian ideal of world cooperation and understanding via science and technology offers hope for humankind in husbanding its remaining natural as well as human resources, and directing their benefits to the satisfying of the needs of everyone, wisely and humanely.

4 Benjamin Henry Latrobe and the Transfer of Technology

Darwin H. Stapleton

The term "transfer of technology" usually means the taking of a technical skill from a nation where it is already established to another nation where it is unknown. Seldom has a transfer of technology been accomplished except by the actual migration, temporary or permanent, of one or more skilled persons. Today, with multiple and complex means of impersonal communication and rapid international travel, this element of the transfer of technology may not be readily apparent, but transfers still depend largely on the migration of skilled persons.

Undoubtedly the most critical era in the transfer of technology to the United States was the period extending from about 1790 to 1850. By the immigration of skilled Europeans and as a result of trips abroad by her own citizens, America kept abreast of the rapid industrialization of Britain and western Europe. Moreover, by the judicious modification of what was imported and substantial contributions of native mechanics and craftspeople, the United States created its own technological tradition, which was soon admired throughout the world. For an understanding of the ways and means of the transfer of technology to America, we must look at the lives of those who participated in the process. One of the most fascinating and significant persons was Benjamin Henry Latrobe, who helped establish the professions of architecture and engineering in America.

Born in Yorkshire in 1764, Latrobe was educated in Moravian schools in England and then Germany. He showed an early aptitude for drawing, and soon took an interest in the allied fields of architecture and engineering. After studying German engineering informally and touring the architectural monuments of western Europe, Latrobe returned to England in 1783. There he probably spent a year or two in the office of John Smeaton, the foremost English engineer, and afterward worked under Samuel

Cockerell, a distinguished architect. In 1791, Latrobe established himself as an independent architect and engineer in London, and thus became a member of two relatively small professions that scarcely existed in the United States.

One of the aspects of English engineering that at this time differentiated it from the rest of the world's was the widespread application of steam engines to pumping water and operating machinery. Smeaton was intimately connected with this innovative steam engine technology, and we may be sure that Latrobe absorbed much of his interest. Latrobe's English career also coincided with the fruitful early years of the engine-building firm of Mathew Boulton and James Watt, and he was aware of their work. England was also the world's leader in transportation improvements, particularly canals and turnpikes. Latrobe probably worked under one of Smeaton's pupils as a divisional engineer of the Basingstoke Canal, a waterway near London, and was a consultant on another canal in Essex. Thus Latrobe was immersed in some of the most recent technical developments of the time.

In architecture, England was the focal point of the classical revival. This school emphasized such elements as the dome, porticoes with Ionic and Doric columns, and a simple, unadorned appearance. This style was the major influence on Latrobe's taste, and the private dwellings in England that he designed reflect it.

Latrobe left England for the United States late in 1795 for personal reasons, probably stemming from his wife's recent death, and arrived at Norfolk, Virginia, in March 1796. For the rest of his life, he remained in the United States and called himself an American.

Latrobe's best-known legacy from his American years is his architecture. While in Virginia he designed his first major work, the Richmond Penitentiary, a massive masonry structure with a plan reflecting the best penological theory of the day. Moving to Philadelphia in 1798, Latrobe had his first opportunities to express his classical revival ideas in the Bank of Pennsylvania, generally thought to be one of the most beautiful public buildings in American architecture, and the Centre Square Engine House, an extremely pure and severe example of the style. After Thomas Jefferson appointed him surveyor of the public buildings of the United States in 1803, Latrobe undertook the redesign and construction of the U.S. Capitol and White House, giving them much of the classic beauty that can be seen by visitors

The Centre Square Engine House of the Philadelphia Water Works exemplifies Latrobe's classical revival architectural style (engraving by W. Birch and Son).
Source: National Archives

today. He also designed many private homes throughout the mid-Atlantic states. In addition to these secular buildings, Latrobe was the architect of several churches, of which the largest and most impressive is the Baltimore Cathedral. The most exquisite is probably St. John's in Washington ("the Church of the Presidents")—a characteristically simple expression of his taste.

As if such a prolific architectural career were not enough for one man, Latrobe was the most active engineer in America during the first two decades of the nineteenth century. Latrobe's greatest engineering commission was the Philadelphia Water Works, which provided America's largest city at

An engraving of Latrobe's design for the U.S. Capitol, published in R. Ackermann's *Repository of Arts* (1825), after an earlier watercolor by Latrobe.

the time with the first comprehensive water system in the United States. He also designed and built the first waterworks in New Orleans. In transportation, Latrobe played a significant role. He directed the navigational improvement of the lower Susquehanna River for the state of Pennsylvania, was the engineer of the Chesapeake and Delaware Canal during the first attempt at its construction as well as the Washington City Canal. He was a consultant on a number of canal projects and turned down positions with others. Latrobe also designed or directed industrial establishments, including America's first steam-powered rolling mill, a forge, and a sawmill and block mill at the Washington Navy Yard, one of the first steamboat yards in Pittsburgh, and a woolen mill in Steubenville, Ohio.

This staggering list of Latrobe's activities shows that he brought a wide range of knowledge and skills to his adopted nation, and further consideration of his projects indicates that there were several ways in which Latrobe transferred his technical knowledge to America. Perhaps the most striking was Latrobe's transfer of technical concepts by making them believable and acceptable to Americans, and persuading people to adopt new ideas. His role in the establishment of the Philadelphia Water Works is a good example of this process.

Although during the 1790s the city of Philadelphia had been troubled by its dependence on contaminated and inadequate wells for water, it had found no plan for remedying the problem that met with the approval of the citizens or city councils. Mechanical waterworks existed in the United States, but nowhere was a system established that was suitable for a major city. On the other hand, urban waterworks were common in Europe, and Latrobe was familiar with them. While he lived in London, he observed particularly the several innovative steam-powered waterworks that supplied the city with river water distributed through wooden pipes under the streets.

Late in 1798, Latrobe came to Philadelphia to erect the Bank of Pennsylvania and attracted the attention of a committee of the city councils that had to look into the water problem. He was asked by the chairman to examine springs north of the city and determine their suitability for a water supply. The result of his investigation was a report that exceeded his mandate and presented his ideas for a comprehensive waterworks based on English precedents. Latrobe's proposal revolutionized Philadelphia's thinking about how to obtain a better water supply. Although he did consider the springs as a source, Latrobe rejected them as insufficient for the purpose of washing the streets and cooling the air, which he and many residents considered the most pressing needs of the city. Latrobe proposed taking water from the adjacent Schuylkill River, pumping it up by two steam engines to an elevated reservoir, and distributing it throughout the city via wooden pipes. He thought that such a system could be ready in seven months.

The most daring element of the plan was the use of steam engines. There were at this time only three steam engines in the United States (none of them in Philadelphia) and no regular steam engine foundry. American knowledge of the capabilities and liabilities of engines was fragmentary. Latrobe attempted to educate the public by reporting all that he could remember of steam-powered waterworks in London and calling on citizens who had visited England to verify his report. He pointedly described the Chelsea Waterworks, one of the earliest waterworks with one of the famous Boulton and Watt engines. To allay fear and skepticism of steam engines, he wrote,

Soon after the[ir] invention, steam engines were justly considered as dangerous, man had not yet learned to control the immense power of steam, and now and then they did a little mischief. A steam engine is, at present, as tame and innocent as a clock.

I have no doubt that this city can produce Smiths capable of constructing very efficient Engines, under proper direction.

The committee of the city councils was taken with the authority of Latrobe's proposal and immediately employed him as consulting engineer. By March 1799, Latrobe contracted for two engines from the Soho Foundry of northern New Jersey, where a former Boulton and Watt engineer was employed. On March 2, the city authorized the construction of the water-works on the basis of Latrobe's estimate of $150,000 for the project.

The critical feature in the city's adoption of this project was Latrobe's informed and enthusiastic presentation of the innovative waterworks concept. His English engineering skills and knowledge permitted him to convince the citizens to solve their water problem by adopting the "latest" technology. In the end, Latrobe's design proved more costly than estimated and was always a drain on the city treasury. In fifteen years it was replaced by a new system, which in seven years was replaced by a third, but in spite of that, Philadelphia remained committed to the concept of a comprehensive urban waterworks and, indeed, was the center of developments in waterworks technology for the first half of the nineteenth century.

Latrobe also brought to America the professional engineering skills of planning, organizing, and directing technical projects, such as canals. Americans had projected many canals since their revolution, but few had been started by the time Latrobe arrived in the United States. One well-trained English engineer, William Weston, had preceded Latrobe, and contributed his skills to canal projects in the middle and northern states, but he only partially trained a few men to continue his work before he returned to England in 1799. Like Weston, Latrobe found Americans interested in and appreciative of his skills in canal engineering. For example, the Washington Canal Company was first organized in 1802, and the directors hired Latrobe to conduct the preliminary survey in the next year. Since he divided his time between public buildings in Washington and preliminary surveys for the Chesapeake and Delaware Canal, he did not write his survey report until early in 1804. When the stock subscription books were opened in June, his report was exhibited as proof of the canal's feasibility. Unfortunately, the canal's promoters were unable to gather enough capital to support the venture.

In 1809, the company was rechartered, and this time a sufficient number of shares were subscribed. The directors again approached Latrobe for advice, but he refused to help unless regularly appointed the canal's engineer. The directors agreed to the arrangement, provided they could deduct from his salary the $300 he had received from his 1804 report. Latrobe acquiesced to the stipulation and also agreed to their condition that he make the canal "on the most oeconomical scale, and in the most oeconomical manner which is consistent with its utility and the provisions of the law." That meant making the locks and other structures of wood instead of stone, Latrobe's favorite construction material. Latrobe in this regard was like virtually all engineers with English training, believing that to build well was to build permanently, and to build permanently required stone. He never forgave the Washington City Canal directors for forcing him to build wooden locks, and was bitterly pleased when the locks deteriorated within two or three years and the company asked him to rebuild one of them as a stone lock.

Latrobe's belief in English professional standards brought him criticism in other circumstances, but in this case he felt vindicated. He wrote,

In our haste to be the most enlightened nation in the world, we forget that our very existence is of recent importation from Europe. ... One part of this error branches out into an opinion that Europeans are expensive merely for the pleasure of being so, that it is a sort of disease incident to their climate as oeconomy is a virtue indigenous in our soils. Oeconomy! The oeconomy of Pendulum mills, & building up the Capital by piecemeal, and canals receiving all the Wash of the land, & wooden locks, & a thousand other cheap things. I am ashamed and angry at it, as an American.

In spite of criticisms of his English engineering attitudes, Latrobe was highly valued as a planner and organizer, and was employed on many technical projects.

Latrobe not only transferred technology in his own person but also aided the transfer of technology by finding jobs for many skilled Europeans and Americans trained by Europeans. Persons competent to supervise works under his direction were particularly valuable to Latrobe because he often had two or more projects under way at once. Thus for the Philadelphia Water Works, he insisted on the right to select a man paid by the city to supervise daily activities and chose John Davis, who had architectural and engineering experience in England. John Lenthall, his man at the U.S. Capitol, also had an English professional background. For his assistant on the Chesapeake and Delaware Canal he selected Robert Brooke, a Philadelphia surveyor who had worked under Weston. Latrobe exhibited the same interest in the English talents of the carpenters, masons, and contractors who actually performed the labor. At the rolling mill he hired John Parkins, an iron worker from Sheffield; at the Philadelphia Water Works he employed Thomas Vickers, a mason who worked under Weston; and on the Chesapeake and Delaware Canal he made a contract with Charles Randall, whom he described as "a practical road and Canal-maker from England."

Latrobe gave many of these skilled people their first American employment, and several went on to long and distinguished careers here. Davis, for example, left the Philadelphia Water Works in 1805 to become the engineer of the first waterworks of Baltimore. Others worked for Latrobe on more than one project or were referred by him to other works. Such a man was Randall, whom Latrobe also employed on the Washington City Canal, and for whom he obtained a contract on the National Road, the federal highway from Cumberland, Maryland, to Wheeling (now West Virginia).

Latrobe promoted the transfer of skills by training a few pupils in his office as well. These were young men whom he took in as apprentices, educated in such basic skills as drafting and surveying, and gave valuable experience in supervising his projects. The most important pupils were Frederick Graff, William Strickland, Robert Mills, and Latrobe's own son, Henry Latrobe. By this personal system of tutelage, which was a common means of training engineers in England, Latrobe transferred his knowledge, skills, and opinions to a new generation of engineers, and ensured his lasting influence on American technology. Graff, for instance, became the most celebrated waterworks engineer in the country during his tenure as superintendent of the Philadelphia Water Works from 1805 to 1847. In those years, Graff was consulted on virtually every major water supply project in the nation.

Finally, Latrobe contributed to the transfer of technology by his participation in a community of American technicians who were European immigrants or had visited Europe. Such men as engineer Robert Fulton, politician and architect Jefferson, chemist and entrepreneur Eric Bollmann, and merchant and papermaker Joshua Gilpin were Latrobe's close friends and frequent correspondents. Their letters to each other are filled with lively discussions of technical innovations.

These friendships led to formal recognition of Latrobe as one of the United States' foremost professional technicians when he was elected to the American Philosophical Society, the first U.S. "scientific" society, and Chemical Society of Philadelphia. Latrobe contributed several papers to the *Proceeedings* of the Philosophical Society, including the first critical essay on American steam engines.

This is not intended to suggest that Latrobe was the fountainhead of early American technology, nor even that he must have a prominent place in the pantheon of great American engineers. But there are a number of insights into American technological development that are revealed by the study of Latrobe's career.

Latrobe provides a good example of how new or "hot" technology is transferred. It is curious that the economies of several technically less developed nations have developed rapidly apparently because they adopted the *latest* technology available from the developed nations rather than simply adopting the standard, tried-and-true technology. This may be accounted for by taking into consideration the enthusiasms that new technology evokes—especially in the trained technician, but in people generally too— leading to the adoption of innovations even when they are at first costly and unprofitable.

Hence the Philadelphia Water Works required the city government to borrow money and increase taxes during the early years of the waterworks' operation, and caused an operating deficit of over $500,000 in its first fourteen years. Yet this patronage brought to Philadelphia a group of mechanics and steam engineers who made that city the first steam capital of the United States and helped Philadelphia begin its nineteenth-century career as one of America's centers of mechanical engineering. Perhaps we should reconsider the notion that it is primarily the expectation of profit that originates and sustains technical innovation.

Another important lesson that we may derive from Latrobe's career is the significance of other trained individuals to the transfer of his technical skills. Latrobe was a highly skilled man, bringing with him new concepts and abilities, but he needed other skilled men to understand and execute them. As gigantic as his achievements may seem, they could not have been accomplished without such close assistants as Lenthall and Davis, such brilliant pupils as Strickland and Mills, or such able craftsmen and contractors as Vickers and Randall. Latrobe's surviving papers reveal dozens of men who were his collaborators, whom he trusted to execute his designs and instructions. Latrobe cannot stand alone as a hero, because his work is the product of many others' active and creative contributions—including many whose names are lost to us. Could we examine all our technologically great men as closely, we would find how much their accomplishments are inseparable from those associated with them.

The American career of Latrobe is a revealing slice of the early years of the nation. It shows the dependence of the United States on Europe, especially England, for technical innovation and the way innovations were transferred to America. We find that innovations were enthusiastically received, and their introduction involved a large number of persons. We have some notion of how new skills and ideas were transmitted to the next generation. A close look at Latrobe's career leaves us more appreciative of the origins of American technical skills.

5 Eli Whitney and the American System of Manufacturing

Merritt Roe Smith

Commonly acknowledged as the "father of American technology," Eli Whitney (1765–1825) towers above other early nineteenth-century inventors as a technological folk hero. Best known is his invention of the cotton gin—a device that revolutionized agriculture and made cotton king in the plantation South. Of equal technical significance is his association with developing methods of interchangeable production—a seminal part of the Industrial Revolution—which became known during the 1850s as the American system of manufacturing. These contributions made Whitney a legend in his time and one of the great men of the modern era.

One thing Whitney never lacked was good publicity. Biographies, popular histories, and textbooks abound with glittering accounts of how he almost single-handedly inaugurated a "dual revolution" in agriculture and industry—a revolution that purportedly made Americans a "people of plenty" and defined the character of their national life. To generations of schoolchildren, Whitney's name became synonymous with the new machine technology—a democratizing force that helped to stabilize America's shaky economy and legitimize the new republic in the eyes of the rest of the world. Yet as often happens in the building of heroes, publicists created a myth larger than the man. The purpose of this chapter is to reevaluate Whitney's role as inventor of the American system, and place him in the larger context of the nineteenth-century technological development.

* * *

The origins of the legend date back to spring 1798, when the debt-ridden Whitney, his cotton gin business on the verge of collapse, approached Secretary of Treasury Oliver Wolcott about manufacturing muskets for the United States. "I am persuaded," he wrote to Wolcott, "that machinery moved by water (and) adapted to this business would greatly reduce the labor and facilitate the manufacture of this article." This suggestion, coupled with the promise of rapid delivery, evidently appealed to the government hovering on the brink of war because Wolcott offered Whitney a contract for ten thousand regulation muskets. Signed on June 21, 1798, the pact called for final delivery by September 30, 1800. The government, for its part, promised to pay Whitney $13.40 a musket and make generous monetary advances at various stages of the contract. Whitney considered the latter provision particularly important. "Bankruptcy and ruin were constantly staring me in the face," he confided to a friend. "By this contract I obtained some thousands of Dollars in advance which saved me from ruin."

Whitney's entry into the arms business undoubtedly was an act of desperation; the threat of "bankruptcy and ruin" drove him into making a rash proposal. The scale of the undertaking was unprecedented. No American armory, not even the large national establishments in Springfield, Massachusetts, and Harpers Ferry, Virginia, had ever produced five thousand arms annually. What is more, Whitney knew little or nothing about arms making. He soon learned that manufacturing firearms was a good deal more complex than building cotton gins. Before a single musket could be produced, he had to outfit a factory, recruit a labor force, and procure the necessary raw materials. Just how hasty his proposal was became evident when he failed to meet his first contract deadline, the delivery of four thousand muskets, on September 30, 1799. With less than a year remaining on the contract, he still had not equipped his armory. Even so, through some astute maneuvering he managed to convince federal authorities to grant one extension after another until he finally completed the contract in January 1809—more than eight years beyond the original deadline.

Whitney successfully forestalled the cancellation of his contract by cultivating an idea that captured the imagination of government officials. In July 1799, he responded to expressions of concern about not fulfilling his obligation by announcing that he was developing a "new principle" of manufacturing that would not only revolutionize the arms industry but also vastly improve the quality of arms delivered to the government. "One of my primary objects," he wrote to Wolcott, "is to form the tools so the tools themselves shall fashion the work and give to every part its just proportion—which when once accomplished, will give expedition, uniformity, and exactness to the whole. ... In short, the tools which I contemplate are similar to an engraving on a copper plate from which may be taken a great number of impressions perceptibly alike." Here was the first clear intimation that Whitney planned to make arms with uniform and even interchangeable parts. The copper plate analogy not only made sense; it convinced Wolcott and his staff that Whitney was indeed on the brink of an important technical discovery.

Having announced his plan, Whitney continued to champion the uniformity concept before members of Congress and other government officials. His most famous demonstration took place in the nation's Capitol early in January 1801, before an audience that included President John Adams and President-elect Thomas Jefferson. Acting with typical aplomb, he showed how he could fit ten different lock mechanisms into the same musket with the use of a simple screwdriver. The intimation was clear: Whitney was making firearms with uniform parts. Those present were astonished. Even Jefferson, who was familiar with the earlier work of Honoré Blanc in France, marveled at the display. "Mr. Whitney," he later wrote to James Monroe, "has invented moulds and machines for making all the pieces of his locks so exactly equal, that take 100 locks to pieces and mingle their parts and the hundred locks may be put together as well by taking the first pieces which come to hand." "This is of importance," he continued, "because out of 10 locks, e.g. disabled for want of different pieces, 9 good locks may be put together without employing a smith." Jefferson clearly recognized the military significance of uniformity.

The 1801 demonstration made a deep impression on America's power elite. Thereafter, Whitney's name became closely identified with the development of interchangeable parts—an association that eliminated further difficulties in getting extensions on his contract and made him one of the most respected men of his era. Suddenly he found himself lionized by public leaders and hailed as the "Artist of his Country." With recognition came respect, not only as an arms maker, but also as a man of affairs whose opinion and advice was sought on a broad range of social and economic issues. A measure of the man's growing popularity could be seen at his Mill Rock armory near New Haven, Connecticut, where scores of curious sightseers flocked to view the inventor and his system at work. Diarists and gazetteers also began to enshrine "Whitneyville" as a monument to American technological genius. Typical was Edward Kendall's *Travels through the*

Northern Parts of the United States—a three-volume work published in 1809
that briefly described Whitney's operations and echoed Jefferson's praise.
During the next four decades, other writers such as Timothy Dwight (1823),
Denison Olmstead (1832), and Henry Howe (1844) embellished the story.
From this literature emerged the full-blown Whitney myth incorporated
into hundreds of nineteenth- and twentieth-century textbooks, and taught
to millions of schoolchildren.

*   *   *

Whitney's reputation as inventor of the American system stood unchallenged until the 1960s, when researchers began to detect serious discrepan-
cies between the written record and extant artifacts. Most damaging was
the discovery that Whitney muskets, including those made under his first
contract, did not have interchangeable parts. The physical evidence was
unmistakable. "In fact," one scholar exclaimed, "in some respects they are
not even approximately interchangeable!" This, coupled with the fact that
individual lock components of Whitney muskets carry special identifying
marks—something that is unnecessary for truly standardized parts—indicates that Whitney must have staged his famous 1801 demonstration with
specimens specially prepared for the occasion. In short, it appears that
Whitney purposely duped government authorities in 1801 and afterward
encouraged the notion that he had successfully developed a system for producing uniform parts.

Is the Whitney legend completely fabricated? If Whitney did not make
arms with interchangeable parts, should he be written off as a charlatan
and dismissed as having played no part in the development of the American system? The answer to both questions is emphatically *no*. Clearly part
of the legend is salvageable, but an accurate understanding of Whitney and
his work requires a reassessment of the man as innovator, advocate, and
symbol. Such an analysis not only addresses problems of technical creativity but also sheds light on the complex social and ideological character of
nineteenth-century American industrial development.

Mechanization was a vital part of the American system, and Whitney
recognized its importance at an early date. Indeed, his initial emphasis on
machine production, alluded to earlier, was instrumental in gaining his first
contract in 1798. Yet curiously enough, a survey of equipment at Whitney's
Mill Rock armory reveals a modest attempt at mechanization. The articles

listed are not nearly as sophisticated as one might expect to find in a pioneering "best practice" establishment.

In September 1801, a few weeks before Whitney made his first delivery of five hundred muskets under the 1798 contract, young Philos Blake wrote home about the machinery he had seen at his uncle's factory. The letter mentioned a drilling machine, boring machine, screw machine, and trip hammer. All four devices were well known to Whitney's contemporaries. The most interesting unit was the screw machine, a device for hollow-milling screw forgings that Whitney may have copied from the Frenchman Blanc. By 1825, the year of his death, Whitney had added to his mechanical repertoire a stamping machine, "nitching" machine for slotting screw heads, pair of cast iron shears, and two polishing machines. In terms of design, none of these machines represented new concepts. Even the famous "Whitney" milling machine, a self-acting device employing toothed cutters capable of producing both plain and irregular surfaces, is now known to have been built not by the master but instead by his nephew, Eli Whitney Blake, around 1827 and probably copied from a similar "straight cutting engine" designed by John H. Hall (1781–1841) of the Harpers Ferry armory.

Mill Rock was a transitional establishment, a halfway point between the traditional craft shop and fully mechanized factory. The physical plant and labor force were never large (the latter averaging around fifty workers), and production rarely exceeded fifteen hundred muskets a year. Compared with other New England armories circa 1820, Mill Rock's machinery was neither plentiful nor particularly distinctive. Whether built on the premises or acquired from others, the designs reflected simplicity and cheapness. In fact, the conspicuous absence of machinery for performing the weightier tasks of gun making reveals that Whitney purchased a large number of components from outside sources. From the beginning he had subcontracted for finished gunstocks and barrels, and he continued the practice into the 1820s. Whitney intentionally confined his operations to the onsite manufacture of lock mechanisms—unquestionably the most complicated part of the musket, but also the one part that could be produced without a large investment in heavy machinery. This strategy fully comported with Whitney's view of Mill Rock as an assembly shop rather than a fully integrated manufacturing establishment. He made his muskets almost entirely by hand methods, and his limited stable of machinery accentuated this orientation.

That labor-intensive methods dominated production at Mill Rock is clearly shown by the large number of files carried on the armory's inventory. Indeed, hand filing represented the touchstone of Whitney's manufacturing system. The filing shop accordingly became the focal point of activity. Visitors invariably commented on the "molds" used by workers at Mill Rock, attributing much of Whitney's success to their widespread adoption. Commonly known as "jigs," these implements consisted of tempered steel patterns that guided armorers in filing down components to their proper dimensions. Their use was not new, even though Whitney claimed to have invented the technique and extended its applications far beyond anything achieved by his contemporaries. Ironically, physical limitations in the jigs themselves prevented him from achieving what he had promised government authorities in 1801: the production of muskets with interchangeable parts. As it was, jig filing allowed Whitney to substitute unskilled for skilled labor in the productive process, thereby effecting substantial cost savings. Herein lay the true significance of his manufacturing system.

A distinguishing feature of industrial development is work specialization through the division of labor. When labor is divided, as in jig filing, individual work assignments become more simplified while the overall production process becomes more complex. The need for coordination and control is therefore more urgent, and this in turn intensifies managerial responsibilities. Further demands are placed on managers when the eventual growth of an enterprise increases the speed and scale of production. Whitney, to his credit, recognized this problem and began to establish procedures for monitoring the work process at Mill Rock as early as 1798. His efforts took two forms: cost analysis and labor regimentation. Both were aimed at improving the efficiency and flow of production, and both constituted important managerial advances.

A shrewd businessman, Whitney knew that proper attention to costs could mean the difference between profit or loss, success or failure. Unlike other private arms makers who simply estimated costs by adding a dollar or so to the going rates for labor and raw materials, Whitney sought to specify costs with precision. During his 1807 contract negotiations with the War Department, for example, he figured interest as well as insurance charges in calculating the price of his product. By 1812, he was beginning to add depreciation allowances for "wear and tear" on his plant and machinery. Coupled with the use of piece-rate accounting, his method of estimating costs placed him in the vanguard of factory masters who were trying to rationalize production through systematic procedures and regularized practice. As a result, Whitney enjoyed lucrative profits from the firearms business while many of his less advanced contemporaries found themselves operating at the margin. This aspect of Whitney's career has received little scholarly attention, but preliminary research indicates that his contributions to management are quite significant and certainly overshadow the mechanical inventions for which he is best but undeservedly remembered.

Whitney's desire for order and discipline applied especially to the people he employed. Like other early factory masters, he styled himself as a "steward of the Lord" whose special calling was spiritual as well as worldly leadership. Seen in this context, Mill Rock became not only a place of production but also a moral gymnasium where "correct habits" of diligence, thrift, and sobriety were inculcated through the systematic control of everyday life.

The Eli Whitney Gun Manufactory at Mill Rock, Connecticut, attracted many visitors to see Whitney's manufacturing and management innovations (painting by William Giles Munson).
Source: Mabel Brady Garvan Collection, Yale University Art Gallery

Such an orientation allowed Whitney to amalgamate the sacred and profane, thereby identifying godliness with productivity and dealing a serious blow to agrarian critics who railed against the moral decadence of industrialization as a threat to republican virtue.

William Giles Munson's painting of the Mill Rock "Gun Manufactory" reflects Whitney's penchant for social control. Everything in the bucolic setting of this painting centers on the neat and orderly armory. Even elements in the landscape—the mill dam, covered bridge, symmetrical arrangement of worker housing, fenced pastures, and rich foliage—convey a sense of balance and stability. Life within the village paralleled this orderly view of the universe. The factory bell called armorers to and from work as clocked time gradually supplanted irregular patterns of toil. Rules punished indolence and encouraged diligence while the strict regimen itself systematically eliminated individuality from the productive process. Above all, prohibitions against drinking, loitering, swearing, and other "improprieties" during working hours eroded preindustrial traditions that had formerly mixed life and labor. Outside the shops, armorers and their families were expected to observe the Sabbath as well as lead moral lives. Those who transgressed the established norms were chastised, fined, and in some cases deprived of credit and housing. Repeat offenders were fired and blacklisted, and thus their chances of finding employment elsewhere in the region were virtually eliminated. These and other sanctions set a pattern of paternalism that would be repeated time and again in nineteenth- and twentieth-century America.

Despite his public charisma, Whitney was not popular with his employees. Armorers chafed under the pressures of working in a closely regimented environment. Constant supervision accompanied by carrot-and-stick methods gave rise to anxieties that vented themselves in labor disturbances and expressions of discontent. Not surprisingly, Mill Rock had one of the highest labor turnover rates in the arms industry. On several occasions, Whitney posted bills for the return of runaway apprentices. Far more frequently, dissatisfied armorers simply left their jobs for more appealing positions elsewhere. Those who remained at Mill Rock continually grumbled about low wages and the rigidity with which foremen enforced the rules. Serious morale problems existed at Mill Rock, and Whitney's inability to attract and retain competent workmen severely hampered his efforts to attain the engineering ideal of uniformity.

Even though Whitney sought to develop new manufacturing and management methods, his efforts fell short of success. Except for a promising attempt at cost analysis, he introduced little that was startlingly new at Mill Rock. From filing jigs to hollow mills, the techniques he used originated with others. Yet his zealous advocacy of uniform production made him one of the key progenitors of the American system. From the start, he not only popularized the uniformity idea but also persuaded politicians to support policies aimed at standardizing the manufacture of military arms. Time and again, he conferred with ordnance and armory officials regarding the preparation of model arms along with the introduction of new techniques. A frequent correspondent was Roswell Lee, a former protégé whose innovative management made the Springfield armory the hub of the American firearms industry. Because of his reputation and influence, Whitney participated in virtually every important government decision bearing on the manufacture of firearms between 1801 and 1825. In June 1815, for instance, he hosted a meeting in New Haven that devised a formal strategy for standardizing the manufacture of military muskets at the national armories. Present were Lee, Benjamin Prescott (Lee's predecessor in Springfield), superintendent James Stubblefield of the Harpers Ferry armory, and Colonel Decius Wadsworth, chief of the U.S. Ordnance Department and one of Whitney's closest friends. Out of this meeting came the decision to introduce a new Model 1816 musket as well as special inspection gauges at the national armories, and eventually distribute them to private contractors. The adoption in 1823 of these gauges, though primitive by modern standards, signaled the beginning of a sustained effort toward the production of truly interchangeable parts—a goal that would ultimately be achieved during the 1840s.

In the final analysis, Whitney's most important contribution to the American system stemmed not from any technical innovations he made but instead from his active promotion of the engineering ideal of uniformity. He was, to be sure, an entrepreneur who saw opportunities for profit and took them, yet he was also an *advocate* who perceived the larger benefits of precision manufacture and worked assiduously to keep the concept before the public eye. Beyond this, his conscious cultivation of the uniformity system had a larger cultural purpose. At a time when the fledgling United States badly needed psychological anchors and a collective social vision, Whitney helped to identify technical progress with political and economic stability. In his mind, technology meant something more than a means of controlling nature and producing goods; it represented a barometer of national development, an accomplishment in which Americans could take pride, and from which republican institutions could draw strength and vitality. Through technology, then, everything associated with the American dream—material abundance, democracy, freedom, perfectability, and destiny—could be achieved. In time Whitney came to symbolize this union. Along with Samuel Slater, Oliver Evans, and other lesser-known individuals, his name stood for not only material progress through technology but republican virtue in an era of developing industrialism too.

*   *   *

At least three major streams of inventive activity converged during the 1820s and 1830s to produce the mechanical synthesis known as the American system. Two centers of this activity—Middletown, Connecticut, and Springfield—were located along the Connecticut River within forty miles of each other and shared a common heritage going back to the earliest Puritan settlements of the seventeenth century. The third center stood near the junction of the Potomac and Shenandoah Rivers in Harpers Ferry, and reflected mores and traditions quite different from those of New England. Yet despite these cultural contrasts, all three communities developed exciting new technologies, built on one another's advances, and eventually succeeded in manufacturing firearms with interchangeable parts. Together they exercised a cumulative impact on American metalworking and shop practice that prevails to this day.

Middletown enjoyed a widespread reputation as the home of a number of distinguished arms makers. Among the most prominent were Henry Aston, Robert and J. D. Johnson, Edward Savage, and Nathan Starr and son. Yet as able as they were, none of these men displayed the mechanical ingenuity and skill of Simeon North (1765–1852).

A quiet, unassuming person, North had a prolific career as an arms maker. He secured his first government contract in 1799 and continued as a private contractor, manufacturing horsemen's pistols, common rifles, and Hall breechloaders, until his death in 1852. Apart from his remarkable longevity, North possessed a talent for technical creativity that made him a trendsetter. At a time when most armorers relied almost exclusively on hand tools to execute their work, he sought to mechanize his operations and adopt precision standards. Success came early. By 1816, he had not only equipped his "Staddle Hill" factory in Middletown with advanced gunstocking and barrel-turning machinery but also had succeeded in manufacturing pistol locks with unmarked interchangeable parts. Equally significant, he had begun to form some of the small lock components with a plain milling machine—the first of its kind in America—thereby effecting significant cost savings in file consumption as well as higher standards of accuracy. Ordinarily these accomplishments should have catapulted North to fame and fortune. But lacking Whitney's flamboyance and charm, he failed to attract public notice. While recognized by peers as an ingenious inventor, he died as he had lived—in relative obscurity.

What North initiated in Middletown, Hall extended in Harpers Ferry. A native of Portland, Maine, Hall received a government contract and moved to Virginia in 1819 to manufacture a breech-loading rifle bearing his name. At the outset he asserted that he would make his rifles with interchangeable parts—a point scoffed at by skeptics but strongly supported by ordnance officials who emphasized the experimental nature of the work. Provided with makeshift facilities at the Harpers Ferry armory, between 1821 and 1827 Hall built an impressive stable of machinery that included water-powered drop forges as well as several types of milling, drilling, and profiling machines. Featuring automatic stop mechanisms, formed cutters, and total metal construction, this equipment represented a significant step forward in mechanical design. Attention to small details, such as the use of balanced pulleys, not only increased the speed and efficiency of operation but also extended engine life. Beyond this, Hall sought to ensure precision by executing work from special bearing points, adopting standard screw threads, and using over sixty-three gauges during the manufacturing process. These innovations enabled him to achieve tolerances of one-thousandth of an inch, and in 1826, he successfully completed the first fully interchangeable weapons ever made in the United States.

Although Hall conducted his operations in a small pilot plant and failed to achieve significant economies of scale, he is nonetheless a pivotal figure in the annals of industry. What was so startlingly new about his work was the extent to which he mechanized his operations and the impressive results he actually achieved. No one at the time Hall went to Harpers Ferry had been able to master the problem of attaining complete interchangeability in firearms. Much of the excitement generated by his work can be traced directly to his success in combining men, machines, and precision-measurement methods into a practical system of production. In 1819, no one—not even the chief of ordnance—was absolutely certain that mechanized processes could be successfully applied to every aspect of gun making. Hall showed that they could, and in doing so, bolstered confidence among arms makers that one day they would achieve in a larger, more efficient manner what he had done on a limited scale. In this sense, Hall's work represented an important extension of the Industrial Revolution in America: a mechanical synthesis so different in degree as to constitute a difference in kind.

Like Hall and North, armorers at the Springfield armory played an indispensable role in fashioning the American system of manufacturing. Under the dynamic and resourceful leadership of Lee, Springfield not only became the "Grand National Armory" but emerged as one of the most progressive manufacturing establishments in the United States as well. Lee recognized that no matter how refined manual skills became, muskets made by hand simply could not compare in accuracy and workmanship with those finished by machinery. Process refinement consequently became the keystone of his policy, and from his appointment as superintendent in 1815 to his death in 1833, he doggedly pursued the mechanized production of uniform parts. Although complete success eluded Lee during his lifetime, he nevertheless established a managerial strategy at Springfield that culminated during the 1840s in the design and manufacture of the U.S. Model 1842 musket. This was the first regular-issue weapon ever to be mass-produced with interchangeable parts, and as such, constitutes a milestone in the history of technology.

Springfield's leadership in the mechanical arts depended less on in-house inventive activity than on an uncanny ability to monitor and assimilate work being done elsewhere. Lee and his successors, John Robb and Major James W. Ripley, excelled at ferreting out technical novelties and incorporating them into the productive process in Springfield. Because the most important private arms makers—like North and Asa Waters—depended almost exclusively on the federal government for contracts, and because ordnance officials insisted that contractors share new technological information with the national armories, Lee and his staff had access to virtually every crucial workshop in New England. Inspectors and machinists from Springfield were continually being dispatched to private armories during the 1820s and 1830s to examine new machinery as well as make drawings so that the devices could be replicated in Springfield. Through this intelligence network, Lee procured invaluable information on the latest techniques of milling, forging, and gauging. Via the same network, he learned about Thomas Blanchard, a young mechanic from Millbury, Massachusetts, who in 1818 devised an ingenious cam-operated lathe for turning the irregular surface of gun barrels. After conducting a demonstration in Springfield, Blanchard not only installed the machine at the national armories but also went on to apply the principle of eccentric turning—tracing a master pattern and reproducing it—to the manufacture of gunstocks. Lee deemed this invention so significant that he subsidized Blanchard by bringing him to Springfield in 1822 as an inside contractor. During the next five years, the Millbury mechanic designed and perfected fourteen different woodworking machines that completely mechanized the process of gun stocking, ensured uniformity, and eliminated the need for skilled labor in one of the three major divisions of armory production. Rarely have the contributions of one person effected such a sweeping change in manufacturing practice in so short a time.

The presence of Blanchard and other gifted mechanics in Springfield helps to explain the armory's surge to international prominence during the pre–Civil War period. As early as 1825, Springfield had become the linchpin of the American arms industry as well as a pivotal clearinghouse for the acquisition and dissemination of technical know-how. If an arms maker or any other manufacturer for that matter wished to learn about the latest developments in metalworking, he invariably visited the national armory. There he could examine the most advanced tools and machinery, compare notes, and discuss common problems with such master machinists as Adonijah Foot, Cyrus Buckland, and Thomas Warner. After satisfying his curiosity and determining his needs, he might then set out for other establishments to gain further knowledge about new and interesting mechanical ideas. The ability of arms makers to move freely from one shop to another contrasted with more closely guarded practices in Europe. Such an "open-door" policy doubtlessly helps to explain the highly integrated character of the American arms industry and the relative speed with which manufacturers assimilated the new machine technology. The same policy also accentuates the public-service orientation of the U.S. military service and armory system during the antebellum period.

By the 1840s, manufacturing techniques aggregated and synthesized in Springfield began to spread to other armories and machine shops throughout the United States. Frequently this influence manifested itself through workmen who had received their early training in Springfield, and subsequently moved to new positions as machinists and supervisors at other manufacturing establishments. These "graduates" of the armory became key disseminators of the new technology and could be found in most of the important metalworking centers in America. In 1842, for instance, Warner, one of the architects of the system in Springfield, resigned his position as master armorer and moved to the Whitney armory in New Haven. Several years later, the Ames Manufacturing Company lured a skilled machinist named Jacob Corey MacFarland to its factory in Chicopee, Massachusetts. Like Warner, MacFarland possessed intimate knowledge of the techniques used in Springfield, and as foreman of the Ames machine shop, introduced "armory practice" in Chicopee. Other manufacturers followed similar strategies, thereby acquiring the latest armory know-how, and relieving themselves of lengthy time lags and costly expenditures associated with learning by doing. Such methods of transfer and convergence carried mechanized production to a heightened level of maturity during the 1840s and 1850s, paving the way for the rise of the American machine tool industry and its entry into European machine markets prior to the Civil War.

The experience of the Ames Manufacturing Company aptly illustrates how private firms availed themselves of the new technology. Established in 1834 by the brothers Nathan P. and James T. Ames, the company became one of the first businesses in the United States to manufacture and market a standard line of machine tools. The firm also undertook special jobbing contracts, and made a wide variety of millwork, mining equipment, cutlery, small arms, cannon, statuary, and other metal castings. Yet despite their obvious technical skill and versatility, the Ames brothers were basically copyists rather than innovators—and for good reason. Chicopee was only a few miles from Springfield, and the brothers had ready access to patterns and drawings owned by the national armory. Aided by a number of former Springfield armorers, their stock of machinery clearly reflected this influence.

Through the marketing activities of Ames and other early machine tool companies, armory practice started to spread to technically related industries, and by the 1850s, it could be found in factories making sewing machines, pocket watches, padlocks, and hand tools. From these modest beginnings it was only a matter of time before the new technology would be adapted to the production of typewriters, agricultural implements, bicycles, cameras, automobiles, and a whole host of products associated with the mass production industries of the twentieth century. Interestingly enough, mechanized techniques as opposed to precision processes constituted the most important transfer from the arms industry. Since precision production was expensive, most businessmen contented themselves with manufacturing highly uniform but not necessarily interchangeable parts. Only the government could afford the luxury of complete interchangeability.

The capstone of two generations of creativity in the American firearms industry came in 1851 at the London Crystal Palace Exhibition. For the first time, little-known Yankee manufacturers impressed visitors with the quality and precision of their machine-made products. Exciting considerable comment and admiration were the unpickable padlocks of Alfred C. Hobbs of New York, Samuel Colt's revolving pistols, and six completely interchangeable rifles made by Robbins and Lawrence, a small "armory practice" company from Windsor, Vermont. All three firms won medals at the exhibition, but more significant, the excellence of their wares prompted the British government to send an investigatory commission, headed by Joseph Whitworth and George Wallis, to New York's Crystal Palace Exhibition and, from there, on a fact-finding tour of the northeastern section of the country during summer 1853. The following spring, Her Majesty's Ordnance Board, eager to take advantage of the American system, dispatched another three-man "Committee on Machinery" to the United States for the express purpose of introducing similar improvements at the Enfield armory near London. By the time the investigators returned home in August 1854, they had placed orders for over $105,000 worth of machinery, chiefly with Robbins and Lawrence as well as the Ames Company. Within four years, several other governments sent similar investigatory teams to the United States. Although little noticed at the time, these events signaled America's coming of age as an industrial power.

Whitney doubtlessly would have been pleased with the outcome. Although he failed to achieve interchangeable standards during his lifetime, he could take considerable pride in the fact that he had been instrumental in popularizing the engineering ideal and fixing the concept in public consciousness. Through Whitney, the uniformity system became a cornerstone of Ordnance Department policy. Through the Ordnance Department, creative mechanics like Hall and North received government contracts enabling them to pursue the elusive goal of interchangeability. And through Hall, North, and other contractors, the new techniques coalesced at the Springfield armory; from Springfield, they filtered through the nascent machine tool industry to permeate American and European markets. Out of this complex tapestry of invention, cooperation, and diffusion can be discerned a mechanical genealogy that directly links Whitney and other early arms makers with the mass production industries of the twentieth century. The achievement sealed the fate of preindustrial culture in America.

Note

Parts of this chapter are adopted from my book, *Harpers Ferry Armory and the New Technology: The Challenge of Change*. I am grateful to Cornell University Press for permission to publish this information.

6 Thomas P. Jones and the Evolution of Technical Education

Bruce Sinclair

There is a long tradition of popular science education in the United States, and it reflects two important tenets of the national faith. One is that technology—science applied to practical purposes—is central to the nation's destiny; the other is that education is crucial to democracy. Few individuals were more clearly identified with these doctrines than Thomas P. Jones, spokesman for a movement of educational reform, editor of the *Journal of the Franklin Institute* from 1826 until 1848, and longtime commentator to the nation on American inventions. In his position as editor of the country's most influential mechanics' magazine, Jones championed the cause of mechanics' institutes and lyceums, apprentices' libraries, and all those similar associations that sprang up in the Jacksonian era to provide useful information to the working classes, so that they might rise to "their proper rank in a republican society."

Although Jones was born in England and educated there as a physician, his varied experiences proved remarkably appropriate to a career promoting the development of American ingenuity. His sentiments for reform were undoubtedly stimulated by an association with Joseph Priestley, the celebrated English scientist, political radical, and religious dissenter, with whom he may have immigrated to the United States around 1794, and perhaps that connection also turned Jones's interests to science. During the next twenty years, he accumulated a considerable reputation for his lectures on various aspects of science and its applications. His talent for lecturing attracted large audiences, and gave him an appreciation of the popular demand for scientific knowledge and some positive ideas about the ways it should be conveyed. In the period before he became editor of the Franklin Institute's magazine, Jones also gained academic experience teaching the principles of science—he was a professor of natural philosophy at the College of William and Mary for four years—and learned the practical skills of the mechanic arts from a partnership in a Philadelphia fire engine manufacturing firm. These activities made Jones well known within the city's community of skilled craftsmen, and when the managers of the Franklin Institute wanted to begin a publication, he was the only person they thought of to edit it. They encouraged him to buy the *American Mechanics' Magazine* from its New York publisher and named him to the professorship of mechanics in the institute as a further inducement to locate the magazine in Philadelphia.

Jones designed the periodical, which he named the *Franklin Journal and American Mechanics' Magazine* in order to be broad in its appeal as well as purposes. He had two great ambitions for his magazine: that it give working-men plain and practical information, and that it direct American inventive activity into constructive channels. This latter dream took on additional potentiality when Jones was appointed head of the U.S. Patent Office in 1828. The connection seemed ideal. Many Americans thought of the Patent Office as a sort of educational agency, and his friends considered Jones perfectly suited to it, both for his learning and practical outlook. Moreover, the Franklin Institute had just assumed ownership of Jones's periodical, and his new official position promised national significance for the journal. Indeed, Jones himself saw the magazine as the way "to lay open those stores of the genius and skill of our countrymen which, although existing in the Patent Office, have hitherto been but very partially known." In other words, the Franklin Institute's journal would also become the Patent Office's journal. To mark the importance of the relationship, the institute appointed Jones editor for life, started a new series of the magazine, and changed its name to the *Journal of the Franklin Institute of the State of Pennsylvania, Devoted to the Mechanic Arts, Manufactures, General Science, and the Recording of American and Other Patented Inventions*.

The editor's official connection to the Patent Office proved to be surprisingly brief, however. After only a year, he was forced out in one of Andrew Jackson's administrative shake-ups. But Jones remained a familiar figure around the Patent Office for another two decades. He conducted a patent agency for a time, was a patent examiner for a short period after the Patent Office was reorganized in 1836, and throughout the years from 1828 to 1848, he brought all his mechanical and journalistic skills to the analysis of new patents for the *Journal of the Franklin Institute*.

The Franklin Institute in Philadelphia was founded in 1824 for the promotion of the "mechanic arts" (steel engraving circa 1830 by Fenner, Sears, and Company from a drawing by C. Burton).
Source: Library of Congress

Jones conducted the magazine with simple editorial rules. He believed that journal articles should be couched in "a style as familiar, and as little technical, as the subject will admit." Plain language was what practical men understood best, he argued; scientific or technical terms quite familiar to an author would still be "dog latin to 99/100 of his readers." Jones was also convinced that variety was essential and articles must be short. Anything longer than eight pages, he claimed, would be ignored by two-thirds of the magazine's subscribers. He had firm ideas about the appropriate contents for his journal, too. Essentially a teacher, his principal aim was to spread knowledge, not to create it. He excerpted material from foreign technical journals, reported on American industrial advances, and encouraged workingmen to share their own insights—but always from the point of view that practical men needed practical information.

Jones was concerned with something different than an effort simply to raise the level of traditional craft practices, however. He was convinced that education was crucial to the social and economic improvement of artisans, "a class of our fellow citizens," he once wrote, "whose importance we are only beginning to recognize." And he also believed that scientific principles lay at the basis of the mechanic arts. If working people would learn those principles, if they could unite natural laws and their practical skills, they could become "scientific" in their work. This new approach to technical education also shared the values of science. It depended on literacy and an open, cooperative style of work, unlike the secrecy that had dominated craft activity for so long. The utility of such a program seemed plain to Jones, and he dedicated his magazine to it. Working people ignorant of first principles wasted time and money on things that would not work. Worse yet, they were misled by mechanical chimeras. As he said in an article on perpetual motion, "There is scarcely any other subject so familiarly spoken of, and so little understood, as the principles of mechanics, and no one, therefore, in which quackery is more certain of success."

Jones brought the same convictions and insights to his analysis of new patents. He envisioned the Patent Office as a great repository of technical wisdom. He saw it, on the one hand, as a museum in which the mechanic could trace the historical progress of the art, and on the other hand, as a collection that described the present state of that art. If properly studied, the materials in the Patent Office would not only save the worker from repeating the mistakes of his predecessors; he would be led to perfect existing processes and devise new procedures. If the mechanic arts could command their history, Jones believed they could also command their future.

Much of the attention Jones gave to patents in the journal's columns was to diffuse the knowledge they contained. The publication of patents was by itself a useful service, since the Patent Office did not begin a journal of its own until 1843. But in his self-appointed role as public commentator on new inventions, Jones provided a range of additional information to the magazine's readers. The specifications of patents he thought particularly interesting were published in full, sometimes with explanatory diagrams and often with further editorial observations. For example, in his list of American patents granted in September 1828, Jones printed the complete specification of a patent for an improved seed planter, including a plate showing the machine and its component parts. In his "Remarks," Jones included the report of a correspondent who had used the planter and his own opinion that the device, developed in New York, would prove especially useful in cotton planting, provided that the seed release mechanism was modified along the lines he then suggested. Thus the mechanic community had called to its attention a mechanically novel device that had been field-tested and might be altered to serve various purposes. Teaching was one of Jones's natural inclinations, and he singled out cases such as the seed planter to illustrate useful and progressive inventive activity. More often, though, he was forced to scold inventors who failed to study the lessons of the past or directed their efforts into irrelevant channels. Of a patent for a revolving belt saw, the editor concluded, "It is mentioned in Rees' Cyclopaedia; has previously been patented more than once in the United States; has been repeatedly tried, and as frequently abandoned as worthless in operation; and such will again be its fate, should it again be essayed by the present patentee."

Patents for steam bedbug-destroying machines, instruments to cure dyspepsia, and in seemingly endless numbers, improved washing machines and butter churns also stimulated Jones to comment. "This is a swinging, or pendulum churn," he said of a combined churn and washing machine, "which really has some novelty about it, and even in this fact alone, there is something cheering, as originality has long been a rare element in churns and washing machines."

What most exasperated Jones were the mechanical frauds, whether or not conceived in ignorance. "This is a most absurd project," he said of Obed

Marston's machine to propel mills, "at least as absurd as the attempt at perpetual motion itself." The specification was also defective, and the patentee had sworn both the oath for citizens and that for aliens. "But even this double fortifying," Jones wrote, "will be of no avail in the attempt to turn dross into gold." Jones used these occasions to stress the importance of scientific principles—natural laws that no jargon could overcome. Yet he was aware that working people were not the only ones deluded in such schemes. By a special act of Congress, Horatio Spafford received a patent in 1832 for "Discoveries in Natural Philosophy, reduced to practice." Spafford's device contradicted the most elementary laws of nature, Jones said, disgusted "that any person having any pretensions to mechanical knowledge, should be, for a moment, deceived by such a plan of obtaining power." In fact, Jones knew that some prominent Philadelphians had invested in the project, and he therefore determined to act in such cases "as a public sentinel."

The journal's monthly list of American patents, "With Remarks and Exemplifications by the Editor," became one of its most popular features. Enlivened by Jones's puckish language and authoritative posture, the column was widely read and extensively quoted. Americans had great expectations for the Patent Office. Mark Twain caught that spirit in *A Connecticut Yankee in King Arthur's Court* when he had Hank Morgan say, "The very first official thing I did, in my administration—and it was on the very first day of it, too—was to start a patent office; for I knew that a country without a patent office and good patent laws was just a crab, and couldn't travel any way but sideways or backwards."

Americans imagined themselves an inventive people; indeed, they saw ingenuity as a national characteristic, and expected the Patent Office to be a monument to that talent. In the self-conscious way they often thought about such things, the patent system would be evidence of their prospects, just as Europe's ancient ruins were testimony to its past. The framers of the Constitution had, from the outset, associated patents with the promotion of "the progress of Science and useful Arts," and thereafter Americans commonly linked invention with the nation's advancement.

Because they used technology to measure their country's development, Americans also made folk heroes of their inventors and elevated technical education to a place of esteem. It became the ideal form of democratic instruction, especially for those who had historically been denied much education from circumstances of birth or occupation. In a sense, useful knowledge became patriotic knowledge, and Jones could argue with sincerity that his magazine helped sustain "an important part of the fabric of our independence."

These cultural values were reinforced by increasing demands from the marketplace for new kinds of technical education. Craft traditions had never provided a successful form of training. Benjamin Franklin's early escape from apprenticeship was typical of the system's failure in America. Instead, a variety of educational devices emerged in the eighteenth century: voluntary self-improvement groups, of which Franklin's Junto was simply the most famous; evening schools that taught applied arithmetic and other practical subjects to urban working people; and at an even more popular level, applied science lectures of the sort Jones had so successfully purveyed. By the early nineteenth century, however, it was becoming clear that this miscellany of technical instruction no longer suited the nation's requirements. Large-scale civil engineering projects to construct canals and railroads as well as complex industrial development that depended on steam power and iron machines called for more sophisticated, systematic engineering education.

Jones was an important figure in this transitional period of technical education in the United States. He perceived two central flaws in the existing educational system. The first was that elementary school instruction was not sufficient to prevent serious differences in educational levels between those who could afford private schooling and those who could not. The second was that apprenticeship systems failed to encourage the intellectual flexibility on which technical progress depended. He thus argued for an approach that linked practical and theoretical instruction to a vision of social mobility. For Jones, and indeed the mechanics' institute movement generally, the solution was to develop new forms of education—low enough in cost to be within the reach of anyone, but without the stigma of public charity schools—more advanced than the elementary schools, yet with an emphasis on science rather than the classics. To achieve these goals, most mechanics' institutes offered inexpensive evening lectures, organized into courses of instruction in the principles of physics and chemistry along with their applications to the mechanic arts. Mechanics' magazines were infused with the same ambitions, and in his journal, Jones experimented with a variety of special features to teach working people a stock of simple but essential scientific principles. And because he assumed that artisans educated in the principles of science would transcend craft boundaries, he also concluded they would transcend the social and economic limits that had historically circumscribed their lives.

In fact, the celebrated union of theory and practice, for which Jones had held such hopes, turned out to have little more than rhetorical value. The principles of science proved unrelated to the concerns of most artisans. They wanted free public education, mechanics' lien laws, and the abolition of debtor's prisons more than the fundamental laws of physics—which in any event had little evident applicability at the level of daily workshop practice. But Jones and his fellow reformers were correct to perceive the need for theory in technical education and necessity of new institutional forms. They were also correct, in an era before the wide establishment of tax-supported public schools, and at a time when private colleges and universities were still dominated by classical traditions, to connect technical education to democratic ideology.

The form that American technical education took embodied both those perceptions. Within a decade after Jones's death, the mechanics' institute movement had been translated into a crusade for university-level training in technology for the sons and daughters of ordinary farmers and mechanics. Also known as the People's College movement, this interest in publicly supported universities of practical science culminated in the Morrill Act of 1862. The act provided the basis for a nationwide system of publicly supported colleges devoted to scientific instruction in agriculture and the mechanic arts—so-called land grant institutions because Congress granted federal lands to the states to maintain the colleges. Indeed, in their mixture of practical purposes, applied science, and democratic aspirations, the land grant colleges gave substance to many of the dreams Jones had for education in his adopted land.

7 Cyrus Hall McCormick and the Mechanization of Agriculture

Carroll Pursell

American agriculture at the beginning of the nineteenth century was largely a handicraft. Only a few tasks, such as plowing and hauling, were widely aided by the power of horse and ox. By the end of the century, many more tasks—most notably those connected with harvesting grass and grains—were done by machines, powered usually by horses or mules, but sometimes driven by steam traction engines or even electricity. The power bottleneck remained—to be broken only after 1900 by the gasoline-powered tractor—yet the machines to mow, rake, and thrash were already at work. An industrial revolution was begun in agriculture, and no small part of the credit was due to the inventor and manufacturer Cyrus Hall McCormick.

As one of the oldest and most fundamental technologies, agriculture had undergone several revolutions over the centuries. The first European settlers in America brought with them a powerful set of devices and techniques that had appeared during the Middle Ages—most notably the horse collar, heavy plow, and three-field system of crop rotation. In the new American environment, they found soil and climate not too different from those of their mother countries, but at the same time they met a set of circumstances that militated against the simple and direct transfer of Old World technology as well as techniques to the New World. For one thing, most of the accessible land was covered with a dense forest that made the plowing of neat furrows through tidy fields an almost-hopeless task. For another, the urgent demand for foodstuffs and export cash crops, coupled with the vast abundance of land as yet uncultivated, made crop rotation schemes, especially those that allowed some fields to remain fallow for periods of time, seem extravagant and unnecessary.

In the short run, the Native American Indians taught the new settlers how to avoid cutting trees by girdling them and how to avoid plowing through the planting of Indian corn in hills rather than furrows. Over the long run, trees were cut down, and as settlement moved westward, the prairies were plowed. Such work was enormously expensive in terms of time and effort. It has been estimated that a farmer in a lifetime of work could hardly hope to clear more than two hundred acres of land, but by 1850, perhaps a hundred million acres had been made ready for the plow.

There was always a shortage of labor on American farms, and the gulf between the task to be done and hands available to accomplish it led here, as in the mills and factories of the nation, to thoughts of mechanization. The idea of using machines to replace people was a typical one in the new nation, where high labor costs and high prices for products coincided. In the northern states of the Union between 1820 and 1870, the rural population increased 2.9 times—but during these same years the urban population increased 14.5 times. The proportion of food sent to the cities rather than consumed in the countryside doubled. New technology in the form of canals, turnpikes, and railroads contributed greatly to these new marketing patterns, and the shift from subsistence to commercial farming in turn stimulated the search for new technology.

Agriculture, of course, was then as now not a single undifferentiated activity but instead a complex, integrated set of tools and techniques, crops, and social arrangements. Wheat, for example, had to be planted on land that was previously prepared, then cut, thrashed, and finally transported and sold. With the availability of horse-drawn plows, a new and widespread transportation network, and a growing commercial market, it was the harvesting (cutting and thrashing) that presented a technological bottleneck to the expansion of wheat culture.

This critical step in the productive process was necessarily seasonal and extremely labor intensive. First the crop was cut and then prepared for drying, usually by being bundled into sheaves to be stood on end in the open field. When dry, these were gathered, and the grain was beaten off the stalk and put into sacks. Bringing in the grain remained what it had been for centuries: a hand process involving cutting with sickles or scythes, and separating the grain from the stalks with hand flails. If this process could be mechanized—that is, if a device could be designed that would substitute horsepower for human toil, the acreage per person could be significantly

Cyrus Hall McCormick
Source: National Archives

expanded. The man who built the machine and solved the problem was McCormick.

McCormick was born on a farm in Virginia on February 15, 1809. His father, Robert McCormick, had for twenty years tinkered with a device to reap wheat, but had never been able to hit on the right design and combination of parts to do the job. At the age of twenty-two, in 1831, Cyrus picked up the task. Rather than modifying his father's machine, he completely redesigned it and combined the seven features of the reaper that today remain standard on all such machines: the divider, reel, straight reciprocating knife to cut the stalks, finger or guards, platform to catch the stalks, main wheel and gearing, and front-side draft traction. His new machine worked reasonably well on local fields during the 1831 harvest, and the following year he ventured to make a public demonstration near Lexington, Kentucky. Still, he persisted in making minor improvements until in 1834, he read of similar work being undertaken by Obed Hussey, who was to be a competitor for many years. McCormick warned Hussey of his priority, and on June 21, 1834, took out a patent for his reaper. In 1843, he began to license the manufacture of his machine to various firms, but the resulting devices often proved to be poorly made, and their failure in the field was frequently attributed to the design itself rather than the workmanship of the particular machine.

In a key decision, McCormick set out to manufacture his own reapers, thus ensuring a quality control that would maintain the reputation of his machine. In 1847, he erected a factory in Chicago, then only beginning the boom that was to make that city the hub of the Midwest. The new location put McCormick in a position to dominate the market of the fast-developing grain belt of the area, and by 1850, he had a flourishing business. Hussey was his first competitor, but after McCormick's basic patent expired in 1848, a horde of new rivals appeared—over a hundred by 1860.

The superiority of McCormick's reaper was revealed dramatically at the first world's fair, the Great Exhibition at the Crystal Palace in London in 1851. At first ridiculed for their paucity and lack of artistry, the American exhibits came eventually to astonish visitors with their practicality and simplicity. On a farm outside London, in wet weather and working a crop not yet ripe, McCormick's reaper beat all other competitors in actual field trials. The London *Times*, no friend of American achievement, generously declared that the introduction of the reaper alone was worth the whole cost of the exhibition. McCormick continued over the next thirty years to win many more major awards in international exhibitions in Paris, Hamburg, Vienna, and other cities.

At home, McCormick proved himself at least as accomplished a businessman as he was an inventor. Subsequent improvements and attachments for the reaper were quickly patented, thereby helping to maintain his market domination. He pioneered in such business techniques as credit selling and mass advertising, and in his factory took advantage of the developing techniques of mass production. By these and other methods he was able to dominate, if not actually monopolize, the reaper market for more than two generations.

McCormick's success was crowned with great wealth and international recognition. He was made an officer of the French Legion of Honor and member of the Academy of Sciences. Much of his fortune was invested in

McCORMICK'S
PATENT
VIRGINIA REAPER.

D. W. BROWN,
OF ASHLAND, OHIO,

Having been duly appointed Agent for the sale of the above valuable labor-saving machine (manufactured by C. H. McCormick & Co., in Chicago, Ill,) for the Counties of Seneca, Sandusky, Erie, Huron, Richland, Ashland and Wayne, would respectfully inform the farmers of those counties, that he is prepared to furnish them with the above Reapers on very liberal terms.
The Wheat portions of the above territory will be visited, and the Agent will be ready to give any information relative to said Reaper, by addressing him at Ashland, Ashland County, Ohio.

Ashland, March, 1850.

A handbill distributed about 1850 advertises McCormick's reaper and illustrates his advertising acumen.

western gold and silver mines as well as railroads, and he was aggressive in seeking to promote international trade for the Mississippi Valley. Some of his fortune went into the endowment of religious institutions, both newspapers and seminaries. In politics he was a Democrat, but his influence suffered because of the decline in the fortunes of that party due to the Civil War.

Despite its enormous success, the McCormick reaper solved only part of the problem of harvesting. The arduous task of thrashing was left untouched by the original machine, and it now became the focus of inventive activity. McCormick himself attacked the problem, and beginning in the 1860s, produced a number of devices to attach to his reaper that would bind sheaves automatically. Many of the earliest thrashing machines were stationary, steam driven or horsepowered, and were positioned in the fields where the cut wheat or other grain was brought to them, and the grain then sacked and carted away.

When the thrasher was combined on a single moving platform with the reaper, the resulting device was known as a combine. During the last decades of the nineteenth century, the vast wheatlands of the Middle Border states (between Texas to the south and North Dakota along the Canadian border) and central valleys of California saw the development and use of huge combines to cope with the enormous farms that were established in

Teams of mules and horses pull large combines in late nineteenth-century wheat fields.
Source: U.S. Department of Agriculture

these areas. Machines pulled by thirty or forty horses or mules cut a swath eighteen feet wide, thrashing, cleaning, and sacking the grain as it moved along. By the end of the century, even larger combines were being pulled by steam traction engines, the first of which had appeared in 1882, just two years before McCormick's death. In 1898, one writer described seeing one "monster of California farming" used for "cutting, thrashing, cleaning, and putting in sacks at the rate of three sacks per minute of barley, each sack weighing one hundred and fifteen pounds, requiring two expert sack sewers to take the grain away from the spout, sew the sacks, and dump them on the ground. Seven men constitute the crew, including engineer and fireman."

The great increase of land put to the plow—the acreage of wheat alone doubled between 1866 and 1878—put a large premium on scale. More power was wanted to work larger machines. As for the past two hundred years, efficiency and a careful husbanding of resources was thought less important than speed and size. By 1890, some 2 million horsepower was being developed by steam on American farms, and that figure had reached its peak of 3.6 million by 1910. Weighing sometimes 20 tons (but rated at only 50 to 100 horsepower), and consuming 3 tons of coal and 3,000 gallons of water per day, these monster machines soon gave way to the cheaper and lighter gasoline-powered tractors. By 1923, there was an average of 4.74

horsepower available for work on each American farm. Of this amount, gasoline tractors provided 17 percent, trucks produced 15 percent, stationary gasoline engines created 14 percent, steam engines supplied only 6 percent, electric motors provided 5 percent, windmills generated 1 percent, and animals still accounted for the largest share at 42 percent. The steady adoption of gasoline tractors and, after 1935 when the U.S. Rural Electrification Administration was established, electricity for the farm effectively provided American farmers for the first time with power cheap as well as and flexible enough to allow a true industrialization of agricultural production.

The appearances of McCormick's reaper, then the combine, and large steam engines to pull them were dramatic and effective steps toward farm mechanization, but for many years they were far from typical. Such machines were so expensive and could work so many acres that only well-to-do farmers with large acreages could afford them, or use them efficiently. The adoption of machines and prime movers varied greatly from place to place and crop to crop. Whole types of farming—such as growing vegetables and fruits, running dairies, and raising chickens—had little use for large steam traction engines and none at all for reapers. Despite untold numbers of ingenious devices from the imaginations of hundreds of inventors, the basic productive unit remained a human being using traditional hand tools. Even in those areas of agriculture where power and machinery were most successful—gathering grain and grass, and tilling the soil—the vast majority of farmers, especially in the northeastern and southern states, were either too poor to afford the new machines, or had farms too small, diversified, rocky, or steeply laid out on hillsides to use them.

For most American farmers therefore, the invention of the reaper was not as important as the far less dramatic improvements in the production of hand tools. By 1850, the American system of manufactures, begun by men like Eli Whitney and John H. Hall in federal and private small arms manufactories, had spread to the fabrication of agricultural implements. In 1849, a Scottish chemist touring the country attended an agricultural fair in Syracuse, New York, and commented that "the general character of the implements was economy in construction and in price, and the exhibition was large and interesting. ... Ploughs, hay-rakes, forks, scythes, and cooking-stoves were very abundant, and many of them well and beautifully made. American ploughs are now exported in considerable numbers. ... The potato grips and forks, of various kinds, cut out of sheet-steel, were very elastic, light, strong, and cheap." Significantly, he remarked that "the cradle-scythes were also excellent: an active man was said to be able to cut four to six acres of wheat a-day with them." He heard news of McCormick's reaper and observed that the "machine, I suppose, must be a good one, from the information here given that as many as fifteen hundred of them have been made at Chicago … this year, and sold for cutting wheat on the prairies of the North-Western States," but the only machine at the fair was one of Hussey's, which people claimed could cut twenty-five acres of wheat a day.

The characterization of agricultural implements as "very elastic, light, strong, and cheap" indicates that they were typical of American technology during these years of the mid-nineteenth century. Two visiting British machine makers noted that "templates and labour-saving tools are used in the manufacture of these implements, which are sold in very large numbers." Plows thus mass-produced by laborsaving machines sold for as little as $2 to $3, making them available to farmers in large numbers.

In a period when most Americans lived in rural areas (it was not until 1920 that a majority of Americans lived in cities), these mass-produced agricultural implements represented a democratic technology at its best. It has been estimated that in 1862, it cost only $968 to equip a midsize general farm, and thanks in part to the spreading techniques of mass production, that cost had fallen to an estimated $785 by 1907. Along with free government land, a cheap, effective, and readily available technology continued to give support to the dream that the nation could remain what it had always been: a country of small, independent, self-reliant agriculturalists.

But that dream, whatever its visionary power, was already doomed. The mechanization so mightily symbolized by McCormick's reaper—with its premium on size and power, and consequent demand for large new sources of energy—was already well under way, and increasingly embraced by those who were well enough placed to take advantage of it and turn it to their own profit. On the one hand, this mechanization allowed fewer workers to feed more people, and thus underwrote the movement toward urbanization and industrialization. On farms themselves, though, the breaking of the harvest bottleneck created new bottlenecks at other points in the productive chain, and hence put a premium on further mechanization. Farmers who before had been able to plow more than they could harvest could now harvest more than they could transport to market. Large-scale farms, worked by large machines, and dependent on huge grain elevators and railroad networks as well as commodity markets, increased the risks and rewards of agriculture. Nonrenewable energy sources of coal and gasoline replaced those of wind and animal muscle. Gradually agriculture began to demonstrate those same characteristics of mechanized, integrated, energy-intensive production that were also coming to mark most manufacturing industries. It was a result that McCormick himself hardly foresaw and for which he was not alone responsible. Yet the reaper, more than any other machine of its century, showed what power and mechanization could do for agriculture.

8 James Buchanan Eads: The Engineer as Entrepreneur

John A. Kouwenhoven

Engineers and architects are grouped together as one of the sixteen classes eligible for election to the Hall of Fame for Great Americans—a national institution established in 1900 and administered by the senate of New York University. Only one person out of the eighty-nine men and women elected to the hall during its first sixty years of existence was chosen in this category: James Buchanan Eads, engineer, who was elected in 1920–1933, years after his death.

Three years before he died in 1887, Eads had been the first American to receive the Albert Medal—awarded from Britain's Royal Society of Arts in previous years to such distinguished Europeans as Michael Faraday, Henry Bessemer, Lord Kelvin, and Louis Pasteur—and again it was for "the services he had rendered to the art of engineering" that he was honored.

Yet in 1871, when Eads was at the height of his engineering career, he was classed as a "Capitalist," along with Daniel Drew, Cornelius Vanderbilt, and Cyrus W. Field, in a popular book of biographical sketches titled *Great Fortunes and How They Were Made* by James D. McCabe Jr., a highly competent journalist and historical writer. And two years after Eads died, Captain E. W. Gould, the garrulous historian of river navigation in America, whose *Fifty Years on the Mississippi* appeared in 1889, and who had known Eads well for many years, claimed that whatever credits were due Eads "as an engineer, or for his mechanical and inventive genius, all sink into insignificance when compared to his ability as a financier."

Gould's judgment was colored, no doubt, by the fact that he had become a large stockholder in one of Eads's ventures and was left with considerable losses after Eads left. Eads had formed a corporation in 1855 to take over the salvage business that had made him wealthy and from which he finally retired in 1857 for reasons of health. Gould, a director and for a while president of the company, preferred to believe that it was Eads's "financial genius" that prompted him to dispose of the company to his friends "just at the moment when the tide of its success was about to turn ... [and] when all that was known of (the) value (of its stock) was from representations of interested parties," rather than because Gould's own bad judgment or bad luck caused his losses.

Yet Gould was nearer the truth about Eads than the authors of two book-length biographies when Gould insisted that Eads's two great engineering achievements—the steel-arch bridge across the Mississippi in Saint Louis that bears his name and jetties that secured a deep channel for shipping at the mouth of the river below New Orleans—were monuments to his financial ability as well as his public spirit and mechanical and engineering genius.

We can make sense of Eads's career as an engineer only if we recognize—as these biographers did not—that he was also a banker, railroad promoter, and one of the creative entrepreneurs who shaped our industrial civilization by finding ways to organize and develop the potentialities of modern technology.

*   *   *

In Lawrenceburg, Indiana, where he was born in 1820, and later in Cincinnati, Ohio, and Louisville, Kentucky, where his father took the family in a series of never-successful attempts to better his fortunes, young Eads displayed the fondness for mechanical things that many boys in western river towns developed in the early steamboat era. By the time he was eleven, he had made a small engine that ran on steam, and after the family moved to Saint Louis in 1833, he had a chance to supplement his native mechanical ingenuity with spare-time reading. He had the use of the private library of a merchant for whom he ran errands and did chores in order to supplement his family's income from the boardinghouse his mother ran. Except for a few years of schooling back in Cincinnati and Louisville, Eads was entirely self-educated.

By the time he was nineteen, his father and mother had moved once more, this time to a new settlement upriver in Iowa, and Eads got his first job on the river, as second clerk (or mud clerk) on Gould's *Knicker-bocker*, running between Cincinnati and Dubuque, Iowa. His mechanical skill had developed to the point where he had designed and built what

Gould described as "a miniature steamboat … ready to raise steam in a tin boiler, ingeniously and systematically arranged." But his job as mud clerk had nothing to do with mechanics; the clerks on riverboats were business managers who dealt with passengers and shippers as well as hired and paid the crew.

Eads quickly demonstrated that he was as talented in dealing with people as he was in engineering mechanical contrivances. When boats were overcrowded, as they often were, or delayed by low water or accident, it was the clerks who were responsible for placating disgruntled passengers and crews. Gould remembered young Eads as one who, "with the suavity that characterized and popularized him in later years, did much to relieve the captain and quiet the irritation, especially in the ladies' cabin."

The *Knickerbocker*, like hundreds of other boats on the Mississippi, was sunk by a snag in 1839. Gould and his clerks managed to get all the passengers and crew safely ashore, but the valuable cargo, including a large quantity of lead from the mines of Galena, Illinois, was lost. River pirates soon made off with the light deck freight, but the tons of lead remained in the sunken hull since there were at that time no wrecking boats equipped to recover it.

The lesson was not lost on Eads. While serving as a clerk on other boats during the next two years, he developed designs for a diving bell capable of recovering the machinery and cargo of wrecked steamboats. Early in 1842, he formed a partnership with the proprietors of the first boatbuilding yard in Saint Louis (established in November 1841) to construct his "Submarine," as he called it. He was ready to go into the salvage (or wrecking) business.

Eads was the active, practical man in the firm, in charge of the boat and salvaging operations (hence the title Captain, which thereafter clung to him), and spent much of his time for the next two years working in his diving bells on wrecks up and down the river. In 1845, however, he married and wanted to be at home in Saint Louis, so he left the salvage firm, and turned his attention to building and managing the first glassworks west of the Ohio River. His unpublished letters, written during the next few years, make it clear that he rapidly mastered the complex technology of flint glass manufacture, learning how to build the melting pots of German clay, mix the sand, pearl ash, lead, and other ingredients, and draw off the molten glass into the pressing molds. But the expenses were great, and the glassworks shut down by the end of 1847. Burdened with debts, Eads went back into the salvaging business with his former partner.

The design of Eads's first submarine—not submersible itself—had been basically an adaptation of the twin-hull snagboats built in the early 1830s by Henry Miller Shreve. Yet Eads equipped his boat with derricks to lower and raise diving bells of his own design along with the machinery necessary to propel the boat, operate the derricks, and supply compressed air to the diving bells. It had worked well, but when Eads returned to the business, he at once began designing and building a series of improved submarines. The fourth, built in 1851 at a boatyard in Paducah, Kentucky, revolutionized the wrecking business, for it was equipped with powerful steam-driven centrifugal pumps designed by Eads under an agreement with the patentee (J. Stuart Gwynne of New York) specifically to clear wrecked steamboats of the heavy load of sand and silt deposited in and on them by the muddy river. Fitted out with these pumps and more powerful derricks, *Submarine No. 4* was able to raise wrecked steamboats entirely and quickly restore them to service instead of merely salvaging their cargoes.

In 1855, Eads and his partner, William Nelson, bought five Shreve snagboats offered for sale by the U.S. government and converted them for use as diving bell boats—one of which, *Submarine No. 7*, the largest and most powerful boat of the kind ever constructed, figured prominently in the next phase of Eads's career during the Civil War.

The government's sale of its snagboats resulted from the decline in federal funds for internal improvements during the period of bitter sectional controversy prior to the war. The work of removing snags and other obstructions from western rivers had almost ceased by 1853, and by 1856, the river commerce of the Mississippi Valley was seriously threatened by the hazards of navigation as well as the inflated costs of insurance on boats and cargoes. In that year, Eads got his friends Congressman Luther M. Kennett and Senator Henry Gever to introduce bills in the U.S. Congress for the improvement of all the major western rivers on a contract basis. At the same time Eads organized a corporation, the Western River Improvement and Wrecking Company, to take over his salvage business. (This was the corporation in which Gould later regretted investing.) It was to bid, as no other organization was equipped to do, for the government contract when Congress passed the bills that his friends had submitted. The House passed its bill early in 1857, but the Senate version died. Similar bills introduced in 1858 were defeated, largely through the influence of Henry Clay and Secretary of War Jefferson Davis.

*   *   *

Meanwhile, Eads had suffered a severe recurrence of the tuberculosis that periodically incapacitated him throughout his adult life and—as indicated earlier—had retired from active work. For the next few years, he and his second wife lived hospitably in a large Italianate villa in suburban Saint Louis, and he invested his considerable wealth in real estate and street railways, and became a prominent director of one of the Saint Louis banks. But even before the outbreak of the Civil War, Eads had worked out a plan for blockading the Mississippi south of Cairo and defending the rivers with ironclad gunboats. Early in 1861, he was called to Washington, D.C., for consultations with the secretaries of war and the navy.

Eads's initial contract with government was to build seven ironclad gunboats in sixty-five days. The boats were not, in fact, completed on schedule, nor did Eads have much to do with their design beyond supplying rough sketches to Samuel Pook, the naval constructor at the Washington Navy Yard who drew up the plans. The delays in their construction were the result of the beleaguered government's failure to make payments as they became due and repeated changes in design ordered by the military authorities. That all seven were completed in a hundred days (during which Eads also converted his *Submarine No. 7* into the ironclad *Benton* to become the flagship of the gunboat fleet) was a triumphant demonstration of Eads's organizational skill. Within two weeks of the signing of his contract, he had four thousand men at work in iron mills, machine shops, and mines from Pennsylvania and Ohio to Wisconsin and Michigan as well as in the foundry and shipyard he had built in Carondelet, just south of Saint Louis, where the keels of four of the ironclads had been laid. The vessels were still unpaid for when they helped capture two Confederate strongholds on the Tennessee and Cumberland Rivers—Forts Henry and Donelson—resulting in the first major Union victories of the war.

While building the ironclads, Eads was thinking about more effective warships for use on the western rivers. Less than a month after the classic 1862 battle between John Ericsson's ironclad *Monitor* and the Confederate *Merrimack*, he was called to Washington by Secretary of the Navy Gideon Welles and asked to design all-iron gunboats of light draft. In the office of the chief naval constructor, he borrowed a drawing board and sketched a design for a single-turreted monitor, turtlebacked from an extremely low freeboard. Eads proposed a turret of his own design, revolving on ball bearings instead of on a central spindle as in Ericsson's patent, and with gunports that closed automatically when the guns recoiled. At first the Department of the Navy resisted Eads's design for the turrets, and the first three iron river monitors he built at his greatly expanded shipyard and ironworks in Carondelet were equipped with Ericsson turrets. But he was subsequently given a contract to build four double-turreted river monitors, two of which were equipped with one Ericsson turret and one of his own design (which became the standard in the future), with its guns manipulated entirely by steam. It was these monitors that served Admiral Farragut so well at the Battle of Mobile Bay.

After the war, Eads continued to design improved gun carriages, turrets, and war vessels. In 1868, after spending a year abroad as a special representative of the secretary of the navy, he submitted a report to the department advocating a comprehensive "System of Naval Defenses." His gun carriage, "run out and otherwise regulated by steam," was exhibited by the Naval Ordnance Bureau in the Government Building at the Centennial Exhibition in 1876, side by side with Ericsson's, which "worked by hand power, taking the united effort of four men to direct its movements." But Eads's contributions to the development of naval armaments were always derogated by the irascible Ericsson (who declined election as a fellow of the American Association for the Advancement of Science because Eads was similarly honored the same year), and they have been largely overlooked by historians. His familiarity with the technology of metalworking and the properties of iron and steel, however, as demonstrated by this naval work, helps to explain his next great achievement.

* * *

Eads had no formal engineering training, nor did he have any experience that seemed, on the surface, to qualify him as a civil engineer capable of designing and supervising the construction of even an ordinary bridge. Yet between 1867 and 1874 he spanned the Mississippi in Saint Louis with a structure of unprecedented boldness. Eads Bridge, as it is named—the only major bridge named for its chief engineer—has justly been called "in many ways the most remarkable engineering job in bridge history." Its masonry piers were founded on bedrock at greater depths than men had ever worked before (or ever worked again for many years), beneath the waters of a river whose violence in the season of high water and when ice jams formed in winter posed problems no earlier bridge builder had encountered. Its arches—two of 502 feet (153 meters), and one in the center of 520 feet (160 meters)—were more than 60 meters longer than any ever built before. It was the first large structure ever erected whose principal structural members (the tubular ribs of its arches) were of steel—a "new" structural material banned for use in bridges by the British Board of Trade until 1877. (It employed not just the Bessemer steel that first became commercially available in the United States in 1865 but also a metallurgically advanced chrome-alloy steel in whose development Eads had played a part.) The novel method employed in the bridge's erection—cantilevering segments of the arches out from adjacent piers until the steel tubes met at the center, thereby avoiding the use of the supporting "falsework" that traditionally obstructed navigation while arches were being built—established a precedent that has been followed in the construction of large metal arches ever since. It also led to the design of the first American railroad bridge of the true cantilever type: the Kentucky River Bridge (1876) by C. Shaler Smith, a young engineer who had been in Saint Louis during the erection of Eads Bridge to design the railroad trestle for its eastern approaches.

Eads Bridge, finished in 1874, pioneered new materials and methods and still carries trains and traffic over the Mississippi River in Saint Louis.
Source: Saint Louis Chamber of Commerce.

Eads was ably assisted by some professionally trained engineers on his staff—including Henry Flad, who planned the method of erection, and Charles Pfeifer, who did the mathematical computation involved in designing the arches, both of whom had been educated and trained in Germany. Eads, though, was personally responsible for the overall conception of the double-decked and triple-arched bridge, detailed design of the pneumatic caissons used in founding its piers on bedrock, and details of its ribbed-arch superstructure. One of the German-trained engineers who worked in Eads's drafting room and was proud of the contributions of his fellow Germans on the engineering staff nevertheless insisted, years later, that Eads deserved "exclusive credit" for designing and proportioning the bridge. "All the minutest details of the tubes with their couplings and pin connections, the skewbacks and anchor bolts, all of it to the last ⅛ of an inch was the work of Eads," he told an audience of engineers.

Eads's improvements on the pioneering European developments in pneumatic caissons included several patented inventions, among them the sand pump, which greatly advanced the technology of underwater excavation. He had familiarized himself with the most advanced European practice before designing his caissons, but unlike other American engineers who contributed to the development of the new technology, Eads had firsthand knowledge of the technical and human problems of working in compressed air, acquired in his diving bells (which are in essence small caissons) during years of salvage work on the bed of the Mississippi. And his courageous determination to use steel as the basic material in his ribbed arches was grounded in firsthand metallurgical experience gained while working with the naval ordnance people during and after the war.

For several months, just after actual construction of the bridge had begun, Eads was so sick that he could not work, and W. Milnor Roberts, an engineer of international reputation and vast experience, temporarily took charge of the project. Said Roberts in a personal letter to Eads after he resumed his duties, "The Bridge, in its inception, in its plan, and in its noble battle against very fierce and extreme opposition, is eminently yours; and on every account you should control every part of its plans and arrangements." But as Roberts also notes, Eads as chief engineer was in a position different from chief engineers in general "not merely from being *projector* as well as the designer of the work" but also because he was "one of the largest owners, and one who has induced the subscriptions, and who is justly looked up to by all the stockholders, and held by them especially responsible."

The story of the financing of the project, and "fierce and extreme opposition" that Eads encountered, as told in Calvin M. Woodward's *History of the St. Louis Bridge* (1881), fully justifies the description of Eads as a creative person who combined the talents of a great engineer with those of a great entrepreneur. Yet not even in Woodward can one find an answer to the question, Why did a man who had never built a bridge before and who had spent most of his life on the river (the role of which as a highway of commerce was threatened by railroads thrusting across its narrower reaches upstream) suddenly marshal all his varied talents to build a railroad and vehicular bridge in Saint Louis, where the combined waters of the Upper Mississippi and two-thousand-mile-long Missouri River presented such formidable hazards?

The oversimplified answer is that Eads had long been involved in several enterprises of great magnitude that his biographers have completely overlooked. When he began planning the bridge, he was the controlling stockholder in the North Missouri Railroad, with one branch building westward to Kansas City, and the other northward to the grain-producing regions of Iowa. As Eads and his associates conceived it, the northern branch would secure to Saint Louis the traffic in grain that had formerly come down the rivers to Saint Louis but was then being reached by railroads thrust west from Chicago and transported on the Great Lakes waterway to the East. The western branch to Kansas City in 1867 seemed to have as good a chance as any of Chicago's lines of becoming a principal link in the first transcontinental railroad system—*if* a bridge in Saint Louis were to join it to the track of the Pennsylvania Railroads' western extensions ending in East Saint Louis. (Tom Scott, the empire-building vice president of the Pennsylvania Railroad, was on the original board of directors of the company that Eads formed to build the bridge, and five of the other eight directors were associates of Eads on the board of the North Missouri Railroad.) In order to finance the completion of the railroad and preliminary work on the bridge, Eads headed a syndicate to buy the State Bank of Missouri—the largest bank in the West—which he converted into a national bank with a million-dollar loan from the nation's largest bank in New York—a loan he secured on his personal note without putting up any collateral.

The financial panic of 1873 and subsequent depression, together with costly delays in completing the bridge caused by political opposition, mechanical and metallurgical problems inherent in developing new technologies, and Eads's refusal to reduce his standards of quality, all made it impossible for Eads to carry his grand railroad scheme to completion. Several years passed after completion of the bridge in 1874 before the tolls received equaled the interest on the bonded indebtedness of the bridge company. The bridge was sold when the bondholders foreclosed their mortgages on it. Meanwhile, the North Missouri Railroad had also been sold under foreclosure, and Eads's bank closed its doors in 1877 in the largest national bank failure up to that time. Two years later in 1879, both the bridge and railroad came under control of financier Jay Gould as keys to his southwestern railroad empire.

*   *   *

By that time, however, Eads had successfully completed the largest project in hydraulic engineering ever undertaken by a private citizen under contract with the U.S. government: the opening of a deep and permanent channel through the bar at the mouth of the Mississippi by constructing jetties extending the banks of the South Pass 2.25 miles (3.6 kilometers) out over the bar to the deep water of the Gulf of Mexico. More than a year before his bridge was completed, Eads had decided that such jetties were the only practicable way to secure an open mouth for the river. The U.S. Army Engineers and Chamber of Commerce of New Orleans thought not. Led by Brigadier General A. A. Humphreys, chief of the Army Engineers, who coauthored the standard work *The Physics and Hydraulics of the Mississippi*, they "proved that such jetties could not be built at the mouth of the Mississippi and would not provide a permanent channel for seagoing vessels even if they could be." Instead, they proposed that the government should build a canal eastward from near New Orleans to Breton Bay on the Gulf.

Relying on his own firsthand knowledge of the river as well as careful study of successful applications of the jetty system at the mouth of the Danube and other European rivers, Eads confidently argued for his jetties before congressional committees. In March 1875, he finally succeeded in getting a bill through Congress authorizing him to proceed according to his plans. The terms of his contract were extraordinary. Only if the depth of water over the bar had been increased in a wide channel from the normal 8 to 20 feet (from 2.44 to 6.09 meters) within thirty months was he to receive any payment whatever. If he did secure such a channel in that time, he would receive $500,000. If he secured 2 additional feet within twelve months thereafter, he would receive another $500,000, and so on. When a channel 250 feet wide and 30 feet deep had been attained, he would receive a total of $4,250,000, with another million coming due over a period of twenty years if the 30-foot channel was maintained.

The $5,250,000 total was considerably less than a congressionally appointed board of civil and military engineers estimated it would cost to build the jetties, and less than half of what a canal would have cost. But Eads had already designed a novel type of "mattress" construction much simpler and more economical than any previously employed, and therefore was confident of making a large profit if he could convince potential investors to help finance the years of work required to earn his first payments.

That he succeeded in raising the money despite a savage campaign of adverse publicity by proponents of the Army Engineers' canal project is further evidence of the "financial genius" that Captain Gould attributed to Eads. That he also devised a novel system of jetty building and all the machines necessary to carry his plans to an early as well as successful conclusion is additional proof of his versatile genius as a practical engineer.

Even before the jetties were completed, Eads had begun planning and organizing what turned out to be his last bold project: a ship railway across the Mexican isthmus of Tehuantepec. This was 1,200 miles closer to the Mississippi's South Pass jetties than Panama, where Ferdinand de Lesseps was laying his disastrous plans to build a tide-level canal. Against the usual opposition of "scientific" and "engineering" authorities, Eads proposed to haul the largest oceangoing vessels overland between the Atlantic and Pacific in huge cradles, drawn on a twelve-rail track by two powerful locomotives, each running on six rails. Once again, none of the government guarantees he asked Congress to provide would become operative until he had built enough of the ship railway to prove that the technological problems had been solved.

A bill to provide a charter and financial aid to his company passed the House in 1880, but died in a Senate committee, and for the next few years similar bills were similarly sidetracked. Finally, ill and desperately weary yet absolutely convinced that his scheme was a national necessity, Eads told Congress that he would finance the entire $75,000,000 project himself, asking nothing of the government but a sanctioning charter for his company.

Just as victory for the charter in both houses of Congress seemed imminent, Eads's health broke, and his doctors ordered him to leave Washington and rest completely. While his wife nursed him in Nassau, his bill passed the Senate, but on March 8, 1887, before it came to a vote in the House of Representatives, he died. His last words, it is said, were, "I cannot die, I have not finished my work."

9 James B. Francis and the Rise of Scientific Technology

Edwin T. Layton Jr.

It has become a truism that we live in an age of science. But the manner in which science influences us has changed. Science used to affect the way men thought; now it affects the way that they live. The work of Nicolaus Copernicus, Galileo, and Isaac Newton changed our view of man's place in the universe. Yet modern science is so abstruse that ordinary people have difficulty understanding it. Thus, while few people understand relativity or quantum theory, everyone is affected by the atomic bomb. The bomb is an example of the interaction of science and technology. Interactions of this sort began in the nineteenth century, in association with the Industrial Revolution. By the early twentieth century, technology had become scientific. It was institutionalized in great industrial research laboratories. It produced a flood of inventions that increased the rate of social change, and altered the way people live and work. The results have not been all good. On the one hand, scientific technology promises a utopia of material abundance, but on the other hand, it threatens nuclear destruction or environmental catastrophe. People differ on how best to direct technical change for human benefit. There is little doubt, though, that scientific technology has become one of the most important forces shaping the modern world.

The rapid rise of the United States from a weak agrarian republic in 1800 to a mighty industrial power in 1900 owed much to scientific technology. Faced with the enormous task of settling a virgin continent, Americans turned eagerly to science. The national government fostered the development of scientific technology in several ways. President Thomas Jefferson founded the U.S. military academy, West Point. Its curriculum was strong in science and engineering, and its graduates were encouraged to become civilian engineers. America's immigration policies also contributed to the development of scientific technology. Among the millions who migrated from Europe to find a better life in the New World, there were a number of highly trained engineers. Engineers, both native born and immigrant, were in the vanguard of those who attempted to base technology on science. They produced many innovations that contributed to America's rapid growth. Among the most important of these men was a young British immigrant, James Bicheno Francis. The hydraulic turbine that he invented became one of the most significant sources of power for American industry. But the tradition of scientific research in industry, which he began, has had an even greater influence.

Francis was born in Southleigh, England. He was trained as an engineer by his father, John Francis, who built an early railroad in his native land. After a brief education at the Wantage Academy, young Francis went to work, first for his father, and then for a succession of canal companies, in the course of which he became thoroughly grounded in the engineering practice of his day. In 1833, at the age of only eighteen, he migrated to America. Competent engineers were scarce in the New World, and Francis became the assistant to one of America's leading engineers, Major George W. Whistler, a graduate of West Point. In the following year, Whistler was called to the industrial city of Lowell, Massachusetts, one of America's first cotton textile-manufacturing centers. Francis accompanied Whistler to Lowell and continued to serve as his assistant until 1837 when Whistler left. The Proprietors of Locks and Canals of Lowell, the corporation that developed the city of Lowell and its waterpower, then made Francis its chief engineer. Thus, after only four years in America and at the age of only twenty-two, Francis assumed one of the most important and responsible engineering positions that America had to offer; for Francis, the American dream of opportunity was fully realized.

Francis quickly vindicated the trust placed in him. His chief duties were to maintain and rebuild a system of dams and canals that provided the waterpower that drove the factories in Lowell. He also had to measure the flow of water to the various cotton mills and advise the factory owners in all engineering matters. In his characteristic manner, Francis patiently gathered data first and then sought rational, scientific solutions to the problems. He found that a half century earlier, there had been a flood before the industrial complex in Lowell had been built. If the water were to rise again to the same height, it would flow over the main canal lock and do a great deal of damage to the new industrial city. Francis therefore increased the height of the canal's sidewalls and built a massive wooden gate twenty-five feet high that could be lowered across the mouth of the canal. Within two years another flood occurred; without Francis's foresight, a large part of Lowell and its factories would have been destroyed, with the probable loss of many lives. As it was, no damage was done, thanks to Francis's scientific approach to his job and good judgment.

Francis depended on the existing scientific literature in engineering, which at this time was mainly European. In some cases—perhaps the majority—this was all that was needed. Following British practice, the factories in Lowell were built on a framework of iron beams and pillars. Francis relied on British data on the strength of iron in evaluating the soundness of plans for new structures in Lowell. But he soon found that in this and other cases the existing scientific knowledge was insufficient, and he had to supplement it by conducting his own experiments. Thus in attempting to use science as the basis of engineering practice, Francis was driven to become a scientist in his own right. This proved to be a congenial role, for he had both keen insight and a flair for experimental investigation.

Francis's career provides an example of how technology became scientific. Engineers borrowed the methods of science, as Francis did in his experiment to determine the strength of iron beams. In this way engineering knowledge, like that of science, became cumulative. Each man's work could build on that of his predecessors. Another trait of science was abstraction, exemplified by its mathematical theory. When applied to technology, this sometimes led to sudden, dramatic changes. The traditional approach to technology had been to make small changes in existing devices. But once engineers understood the theory underlying a device, it was sometimes possible to make rather-radical improvements. A case in point was the invention of the hydraulic turbine.

For centuries, waterwheels had been the basic source of industrial power, but the conventional waterwheel had a practical limit in its efficiency. At best, only about two-thirds of the energy of falling water could be transformed into mechanical energy to drive machinery. By using mathematical theory, the French engineer Benoît Fourneyron produced a radical improvement in the 1820s: the hydraulic turbine. This innovation increased the efficiency of water motors to nearly 80 percent. Francis then improved significantly on Fourneyron's design; indeed, the Francis turbine is still the most widely used type. But Francis also made improvements in the science applied to the turbine that were of equal or greater importance.

To understand what Francis accomplished, it is necessary first to understand what Fourneyron did. France had been a pioneer in attempting to base engineering on science. French engineers, notably Jean Charles Borda and Lazare Carnot, had shown the major limitations of traditional waterwheels, and they had enunciated the conditions for greater efficiency. They found that the energy of the water was lost in two ways: by "shock" at entrance, and by excess velocity at exit. By shock, they meant the turbulent motion of the water produced when a stream of water struck the paddle of a waterwheel. Thus the conditions for highest efficiency were that the water should enter the wheel without shock and leave it with only the minimum velocity needed to clear the wheel.

Fourneyron's achievement was to show how these conditions could be approximated in practice. His turbine was much like a conventional waterwheel turned on its side. The water entered at the center and flowed outward in a horizontal direction. The wheel was enclosed in metal plates to prevent the water from leaking out, and the blades of the wheel were curved. By curving the blades, Fourneyron showed how to achieve the second condition: exit with minimum water velocity. His analysis demonstrated that this condition would be achieved when the water left in a direction precisely opposite to the wheel's motion—that is, in the direction of a tangent drawn to the circumference of the wheel. Fourneyron could approximate this condition by giving the blades of his wheel just the right amount of curvature.

Shockless entry was the hardest condition to meet. This implied that there would be no impact of the water on the wheel. Fourneyron's theory indicated how this could be achieved. The water should enter in a direction exactly tangent to the moving blade; since the water would be moving parallel to the axis of the blade, there would be no impact, and consequently no loss through shock or turbulence. To meet this condition, it was necessary to control the angle at which the water entered the wheel. Fourneyron approximated this by making his wheel hollow and mounting fixed guide vanes inside the moving wheel. These vanes directed the water to the wheel at the required angle.

Fourneyron's turbine represented a remarkable advance. But it was considerably more expensive than traditional waterwheels, and its use was therefore limited to great industrial centers where there was plenty of capital and a pressing need for higher efficiency. A cheaper turbine would be of enormous benefit, particularly in America, where waterpower was abundant but capital was scarce. The Francis turbine, when fully developed, provided the ideal answer; it was much cheaper than Fourneyron turbines and usually got significantly higher efficiencies. Francis's improvements, like Fourneyron's own work, were based on a scientific analysis of the problem. In effect, he expanded the science of the turbine. Yet he did more, for the defects he saw in the theory of the turbine were in fact quite general: they applied to virtually all attempts to apply science to technology.

There were two basic defects in existing scientific technology. First, the mathematical theory was too idealized or simplified. It left out such things as friction, which was of vital importance. Second, attempts to apply experiments to machines had run into perplexing problems due to "scale effects." The results were often not consistent, varying in unpredictable ways as the size of the machine changed. Francis did not solve either of these problems completely, but he found ways of circumventing both limitations. Thus Francis showed not only how to improve the turbine but also how to create a more scientific technology.

Francis was not alone in this work. He collaborated with another engineer, Uriah A. Boyden. It was Boyden who first introduced the Fourneyron turbine to Lowell in 1844, and he patented a number of improvements. A few years later, he sold his patent rights to the Proprietors of Locks and Canals of Lowell, and then collaborated with Francis over many years in improving the design of turbines and methods used in testing them.

The talents of the two men were complementary. Boyden was the abler at mathematical theory, while Francis was the greater experimentalist. Boyden was a shy genius, almost a recluse; he published none of his findings, and had no close professional associates apart from Francis. Francis, who was warm and gregarious by nature, published the results of their collaboration and trained a generation of younger engineers in their methods. Boyden had a profound understanding of the defects of the existing hydraulic science and did much to set Francis on the right course. But Francis had the experimental skills to show how these deficiencies might be overcome. The insights of Boyden and Francis led each of them to see how a radical improvement in turbine design could be made. Francis was apparently first, however, and he was the one to publish his results. As such, the Francis

The Boyden outward-flow turbine is Boyden's modification of the Fourneyron design. The water enters fixed guide vanes at L and moves outward through curved buckets.
Source: James B. Francis, *Lowell Hydraulic Experiments* (Boston: Little, Brown and Company, 1835), plate I.

turbine rightfully bears his name. On these foundations, both scientific and practical, an enduring tradition of industrial research grew up in America that drew inspiration from Francis.

To understand the work of Francis and Boyden, it is necessary to understand the theoretical and practical problems that they surmounted. The existing mathematical theory was beautiful and elegant, but it was too idealized. This defect was a matter of necessity, though. The most important factors omitted from the theory were friction and the internal resistance of fluid. Not until the twentieth century were satisfactory theories evolved to understand these phenomena.* Boyden saw this problem with great clarity and stressed the limitations of idealized theory in his correspondence with Francis. But Francis fully grasped the point and used it. The core of hydraulic science lay in measurements of the flow of water. Yet none of the formulas available were entirely accurate. As Francis wrote after reviewing existing theories,

The result, however, of these numerous labors is far from satisfactory to the practical engineer. On a careful review of all that has been done, he finds that the rules given for his use, are founded on the single natural law governing the velocity of fluids, known as the theorem of Torricelli; omitting in consequence of the extreme complexity of the subject, all considerations of many other circumstances, which it is well known, materially affect the flow of water though orifices.

The existing mathematical theories of the turbine, such as Fourneyron's, "solved" the problem by a convenient assumption: that friction and fluid resistance could be omitted. This was a necessary assumption at that time, if any mathematical theory was to be constructed. But Fourneyron and his successors converted a mathematical necessity into a principle of design. They assumed that the actual path of the water was of no significance as long as there were no sharp bends to produce turbulence. Boyden realized that this assumption was incorrect; eddies and turbulence were produced by many things, not just by sharp bends. And these were among the most important sources of lost efficiency in turbines. Since the existing mathematical theory could not cope with eddies and turbulence, Boyden's response was to abandon this approach while continuing to use its key results. He traced the path of a stream of water particles, the "streamline," inch by inch as it moved through the turbine, and by utilizing basic hydraulic and mechanical principles, was able to take additional factors into account and thus reduce losses of efficiency. His turbines got efficiencies of as high as 88 percent. Francis absorbed and expanded on Boyden's methods, developing graphic methods for designing turbines.

---

* Both friction and water resistance represent a loss of the energy available to do work. Water resistance arises when some of the particles of water move in a direction different than that of the fluid flow. If the motion of the particles is circular, we speak of eddys; if the motion is random, in all directions, we speak of turbulence. While we still lack a comprehensive theory of turbulence, many of the phenomena of fluid motion that perplexed nineteenth-century engineers have been clarified by the development of the science of fluid mechanics in the twentieth century.

This detailed attention to the actual path of the water flowing through the turbine revealed a fundamental limitation in Fourneyron's design and suggested a possible remedy. In an outward-flow turbine such as that of Fourneyron, the inner circumference of the wheel is of necessity less than the outer circumference. Thus the channels through which the water flows get wider as the water moves through the turbine. This causes the water to spread out to fill the widening channels, producing eddying, a circular motion in the water, leading to a loss of efficiency. That is, an outward-flow turbine, of necessity, loses some of the energy available by converting it into internal motions within the water. There was an obvious remedy: to reverse the flow from outward to inward. This had already been done by an American millwright, Samuel Howd. Howd's wheels had the advantage of being compact and consequently less expensive. And an inward-flow wheel eliminated losses through eddying, since while moving inward, the channels got narrower rather than wider. But as is often the case, when one problem was eliminated, another took its place. Narrowing the channel through which a fluid flowed increased its velocity. This was most undesirable, since it meant that the water would have a considerable velocity on its exit from the turbine, representing a serious loss of efficiency. Hence, Howd's turbine was cheap but quite inefficient.

Francis saw how he could combine the cheapness of Howd's inward-flow turbine with the high efficiency of Fourneyron's outward-flow turbine. The basic problem was to slow the water down. In any inward-flow design, the width of the channel must narrow. Francis saw that the critical fact was not the width but rather the cross-sectional area of the channel. So that while the width in an inward-flow turbine must narrow, this could be compensated for by increasing the depth, the distance between the top and bottom plates that enclosed the wheel, which were called "crowns." By altering the depth, the cross-sectional area—and consequently the velocity—could be controlled. Such a turbine is called a "mixed-flow" turbine, since the water moves not only inward but downward as well.

Francis mentioned the idea to Boyden, who shortly thereafter came up with his own version of the mixed-flow turbine. On April 15, 1848, Francis wrote the following memorandum, probably to preserve his own priority:

I met Mr. Boyden on the street. I asked him if he did not think that the making of the distance between the crowns of the wheel greater on the inside periphery than on the outside was new. ... He replied that he had application then pending for patents

Fig. 1.

In this design for the Francis inward-flow turbine, the water enters the fixed guide vanes at E, and flows inward and downward through the buckets. A Prony friction brake, part of a Prony dynamometer, is shown mounted on the upper end of the shaft at 0.

Source: James B. Francis, *Lowell Hydraulic Experiments* (Boston: Little, Brown and Company, 1835), plate VIII.

for this idea. ... I was much surprised—the more so as I had explained to him some months ago that I intended to make the wheels for the Boott Mill with a greater height between the crowns on the inside periphery than on the outside—and the impression I got at the time was that he thought it would not answer and he gave me not the slightest intimation that he claimed it as his invention.

It is probably impossible to determine whether Boyden or Francis had priority in inventing the mixed-flow turbine. Francis evidently thought he did, and apparently suspected that Boyden may have taken the idea from him. But it is also possible that Boyden, who was secretive by nature, had been thinking along similar lines and chose not to tell Francis of his work in order to get a prior patent claim. Boyden's own patents on this type of

This testing flume with Emerson's dynamometer was constructed by the Swain Company in 1869 and is similar to Emerson's flume in Holyoke, Massachusetts, which became a national laboratory for testing turbines.
Source: Journal of the Franklin Institute, 1870, plate I.

turbine represent a rather complex and sophisticated set of ideas, and it is clear that they are not just an imitation. Indeed, they appear to represent the culmination of a great deal of thought over a considerable period of time. It is possible that Boyden's thinking was crystalized by Francis's communication. There is no question, though, that Francis's work was independent, or that he was the first to publish his work.

The Francis turbine was not an instant success. The advantages that it seemed to offer in principle were not immediately realized in practice. This is a rather common situation in technology, which necessitates a lengthy and expensive process of development. Before Francis, inventors had to rely principally on intuition and experience in trying to perfect their devices. Yet in the case of the Francis turbine, the development was made more scientific by the systematic use of experimental testing. Thus theories or ideas of how to improve a turbine's performance could be put to rigorous, quantitative test. The basic tool was already at hand, the Prony dynamometer, which had been developed in France, and was used by Fourneyron to measure the rotary force or "torque" developed by his turbines. It consisted of a massive wooden friction brake attached to a balance arm, on the end of which weights could be mounted to measure the torque in a manner quite analogous to the way in which objects are weighed on a scale. But while the Prony brake could measure the torque or output of the turbine, there was no equally satisfactory way of measuring the input—the amount of water used in a given time. Only by comparing output and input precisely could the efficiency of the turbine be accurately determined. The existing formulas for calculating the flow of water were based on small-scale experiments. Here was a case, however, where the results varied with the dimensions of the turbine as well as the gate or weir over which the water flowed.

Among Francis's greatest achievements was the development of turbine testing into a matter of exact science. In this, too, he followed Boyden, who perfected several improvements in testing methods and apparatus. To compensate for the scale effects, Boyden and Francis did all their experimentation on full-size turbines, thereby circumventing the unpredictable results of model experiments. But Francis went further, and determined a new and more accurate formula for calculating the flow of water over weirs. In this he got some mathematical assistance from Boyden. The formula was basically empirical, though, and it was Francis who conducted an exhaustive set of experiments to determine the new formula. Francis then applied it, along with other improvements, to testing turbines with a degree of precision never before attained. Perhaps most important of all, he published his findings in 1855 in his monumental book *Lowell Hydraulic Experiments*. This book became a guide for American turbine builders in improving and testing the turbine. Francis also trained a number of assistants in his methods, and some of them went on to test turbines. A number of companies conducted their own tests, but in 1870, a former employee in Lowell, James B. Emerson, organized the Holyoke Testing Flume, where he conducted public tests of turbines and published the results. The flume was, in effect, an industrial research laboratory. The result of scientific testing was a spectacular improvement in the performance of Francis turbines and an even more remarkable decline in their cost. By the 1870s, variants of the Francis turbine became the most widely used hydraulic prime movers in America, at a time when waterpower was still more important than steam for industrial purposes. Yet however significant the Francis turbine, Francis's methods and insights were of even greater significance in the long run. Building on the work of European forerunners and guided by Boyden's insights, Francis showed how a truly scientific technology could be constructed. He developed theoretical and experimental methods that made turbine design more of a science and less of an art. He showed others how to use his methods and insights, and his followers carried on a tradition of industrial research that finds expression in the great industrial research laboratories of the present day. Francis, of course, was not alone. Other engineers in other fields of endeavor were following similar paths. And it was the confluence of these paths that created, for good or ill, modern scientific technology.

10 Alexander Graham Bell and the Conquest of Solitude

Robert V. Bruce

The Industrial Revolution, begun in the eighteenth century, led in the last half of the nineteenth century to a communications revolution that shaped the course of industrial and urban growth: modern industrial society could not exist without the instant interchange made possible by developments in communications. For much of the world, the telephone is the symbol of the communications revolution. It not only serves the office and factory but also has entered the home to form a major social and personal link between people. Today's telephone systems are global and expanding as fast as installation will permit in most countries. The United States is a nation of telephones with a total of 162 million as of 1977, and with 96 percent of its homes having telephone service.

The search to transmit speech electrically was part of the burst of individual inventive activity that swept nineteenth-century Europe and America. The invention of the telephone was almost inevitable with the number of minds bent on speeding communications and the accumulating advances in science, and in the 1870s, a Scottish-born teacher of deaf-mutes, Alexander Graham Bell, decided to make his fortune by inventing an improved telegraph and came up with the telephone as well.

Appropriately, Bell came from a family preoccupied with speech: his grandfather had taught speech and elocution, and his father, Melville Bell, also made speech his career. At the time of Alexander's birth in 1847, his father was well established in Edinburgh teaching speech, writing, and lecturing on elocution, and beginning his research on the workings of the human voice. In his mother, Alexander found another perspective on the human need to communicate. Eliza Bell was almost deaf, and it was Alexander, the second of the Bell sons, who was able to communicate best with her. By speaking in a low voice close to her forehead, he could make his mother hear without her ear tube. Like most of his family, he had a remarkably expressive, flexible, and resonant voice, but in this peculiarly specialized sort of communication, he seemed better than any of them.

Although he was only a fair student at Royal High School, the young Bell showed a real aptitude for his father's work. He had a keen ear, and while not possessed of absolute pitch, was able to distinguish fine differences between sounds. Moreover, he was a good musician and even considered becoming a concert pianist.

Edinburgh was in the forefront of mid-nineteenth-century scientific and technological developments. Growing up around its workshops and mills, Bell gained knowledge of mechanics, devising a mechanical grain husker, and soon formed an ambition to invent. This, he perceived from the technologically minded men around him, was a way to get ahead. On a challenge from his father, he and his older brother built a speaking machine at home, making gutta-percha replicas of the mouth, throat, and nose, a maneuverable tongue, and bellow lungs, and succeeded in producing humanlike cries.

Melville Bell's speech research had as its goal a complete and universally applicable phonetic alphabet: a written system in which symbols represented the positions of the vocal organs and indicated the articulation of all possible sounds. In 1864, he completed his "Visible Speech" system, and the three Bell sons gave demonstrations using Visible Speech to pronounce correctly sounds along with foreign and dialect words unknown to them. George Bernard Shaw, a family friend, immortalized Visible Speech in *Pygmalion* with Professor Higgins using it to transcribe Eliza's Cockney dialect. Designed primarily for linguistic scholars and phoneticians, Visible Speech also had practical application in teaching the deaf to articulate, since it showed them graphically how to form sounds they had never heard.

Bell's parents moved to London in 1865, where his younger brother subsequently died of tuberculosis. After teaching in Elgin and Bath, Bell joined the family in 1868 to enter the University of London and help in his father's work. Making use of the potential of Visible Speech, Bell became involved with his first deaf pupils. But before he could complete his studies, his remaining brother died. Already committed to a lecture tour in the United States, Melville and Eliza Bell decided to emigrate to Canada, and convinced Alexander to join them. In August 1870, the Bells settled outside Brantford, Ontario. The following April, Alexander found himself, through his father's connections, with a short-term appointment to teach Visible Speech at the Boston School for Deaf Mutes. He found enough pupils to continue the next year, and in 1873, accepted a professorship of "Vocal Physiology and Elocution" at the recently opened Boston University.

For the rest of his life Bell considered his profession to be a teacher of the deaf, and always so identified himself on biographical forms and questionnaires. And in spite of his many later interests and commitments, he always made time to see parents with deaf children, worked energetically for the education of the deaf, contributed to research, and undertook many individual acts of charity to deaf persons. As one to whom Bell gave personal and financial support in her struggle to overcome both deafness and blindness, Helen Keller dedicated *The Story of My Life* to him.

The Boston area in the mid-1800s hummed with scientific and technological activity. The center of industrial New England as well as the site of Harvard University and the Massachusetts Institute of Technology, the community drew together scientists, engineers, inventors, skilled mechanics and artisans, and businessmen eager to invest their capital. Bell attended many public lectures, read extensively on physics—acoustics and electricity especially—and began a line of work that eventually led him to the concept of transmitting speech electrically.

The scientific principles basic to the telephone had been discovered forty-five years or more before Bell invented it. Danish scientist Hans Oersted had demonstrated the principle of electromagnetism in 1820; England's Michael Faraday had published his work on the induction of current in 1831. Oersted discovered that a current passed through a wire coiled around an iron core magnetizes the core. In retrospect, it seems obvious that if the magnet were brought close to a flexible metal diaphragm and a varying current were passed through the coil, the magnet would cause the diaphragm to vibrate to produce sound. To achieve current variations subtle enough to convey speech, Bell was to use a principle of induction: a changing magnetic field will induce or generate a current in a circuit. The mechanical power of the voice could be used, Bell discovered in July 1875, to vibrate a diaphragm fixed near an electromagnet, changing the magnetic field and thus inducing the necessary varying current. He had the original inspiration in summer 1874 while visiting his parents in Brantford. Bell was later to joke that the telephone was conceived in Brantford and born in Boston, thereby sidestepping the argument over where the telephone had been invented. His first concept was of a series of tuned reeds rather than diaphragms to transmit and receive the vibrations of the voice, but Bell did not believe that his idea would produce audible sounds.

At that time Bell was at work on two earlier conceptions: a "harmonic telegraph" to send multiple messages simultaneously, and a technologically related device he called the "autograph telegraph" to transmit whole documents in facsimile.

Since Samuel F. B. Morse completed his first telegraph line in 1843, telegraphy had burgeoned into a full-fledged technology. Dominated by the Western Union Telegraph Company, this new industry meant instantaneous communication between faraway points that previously had to depend on the physical delivery of messages. The new system, however, still relied on hand delivery between telegraph stations and individuals. Taking their cue from the success of telegraphy, such would-be inventors as young Thomas Edison and Bell looked to improvements in the telegraph to make their fortunes. Bell's telegraphy efforts led him to the telephone and a system that would eventually supplant the telegraph. Ironically, Western Union turned down a chance to buy all of Bell's telephone rights for $100,000 in 1876. Committed to a system incompatible with the telephone, the company shortsightedly missed its opportunity. Another twist of fate later saw the Bell Telephone Company pitted against Western Union in the great patent controversy that developed over the telephone.

Bell drew on his musician's observations for the harmonic telegraph, using the principle of sympathetic vibration—the phenomenon that causes the strings of a piano to resonate selectively, echoing only those notes that are sung into it. He used a set of steel "reeds," each tuned to a different frequency, to open and close electric circuits in vibration, thus producing intermittent currents of corresponding frequencies. The currents could be sent simultaneously over the same wire to electromagnets bearing on similarly tuned reeds. Each receiving reed would vibrate selectively in response to the corresponding reed at the transmitting end, thereby separating out the frequencies. Bell devised a stylus mechanism to be attached to each receiving reed, which would mark a passing strip of paper whenever the reed sounded.

Bell's nascent ideas on the telephone were temporarily pushed aside by his work to perfect his harmonic and autograph telegraph devices. He was spurred on by the news that an electrician and inventor, Elisha Gray,

A model of the first telephone receiver to carry the sounds of speech, set up overnight on June 2–3, 1875, by Alexander Graham Bell's assistant, James Watson.
Source: American Telephone and Telegraph Company

was working on a "musical telegraph." But early in June 1875, Bell discovered by accident in Boston that the theory he had conceived in Brantford would work. He found that plucking the reeds of a harmonic telegraph without any transmitter current would induce an undulating current and the receiver reeds would not only vibrate but also sound audibly. Then, using a membrane diaphragm with a metallic center instead of a reed, Bell transmitted the sound of his voice to his assistant, James Watson, who recognized it but could not make out the words. Bell's next breakthrough came in January 1876. Working on his telephone patent application, Bell added the concept of variable resistance to his specifications, noting that a wire vibrated by the voice while partially immersed in mercury or some other conducting liquid could be placed in a circuit to vary its resistance and produce an undulating current. This meant that transmission no longer had to depend on the limited mechanical power of the voice to induce current, since through variable resistance, an existing current of any strength could be modulated.

But it was not until March 10 that using a variable resistance model, Bell called out, "Mr. Watson, come here. I want to see you." And Watson in the next room heard.

Meanwhile, Bell had placed a completed patent application in the hands of Gardiner Greene Hubbard, one of his two financial backers and his future father-in-law, who in turn had deposited it informally with the Patent Office in Washington, D.C. For reasons involving negotiations with British investors, Bell had told Hubbard to withhold official filing. Hubbard became impatient, and on the morning of February 14, 1876, without Bell's permission, he filed it formally. In a remarkable coincidence, Gray appeared at the Patent Office several hours later to file a caveat—a statement of a conception not yet reduced to practice—for a telephone device.

It seems probable that Gray had heard of Bell's device to transmit speech electrically, since Gray had been in and out of the Patent Office for several weeks before February 14. At times the most important thing in developing an invention is to know that it is possible. In all likelihood, Gray had thus been given the cue to try to do it himself. There was a subtle but fundamental difference between Bell's and Gray's approaches, however, which indicates that neither knew the details of the other's thinking.

There was obviously a conflict, but since Bell had priority in time—although only by a few hours—and since he had reached a much more advanced stage of invention, he was granted the patent—No. 174,465—perhaps the single most valuable patent in history.

Western Union subsequently bought out Gray's rights and hired Edison to come up with patentable improvements to the telephone. Edison developed a transmitter superior to Bell's, and with it, a competing telephone company backed by Western Union began to draw customers away from the Bell Company. In the ensuing year of litigation, it was Bell's patents along with Bell's clear and convincing testimony as a witness that saved the day. In a settlement, Western Union gave up its telephone rights including the Edison transmitter patent for 20 percent of the telephone rental income for the seventeen years that the patents were valid. The Bell Company emerged with a monopoly, which it continued to hold despite some six hundred legal challenges over the next two decades.

The immediate struggle Bell faced after receiving the patent, though, was to have people accept his invention. The Centennial Exhibition in Philadelphia presented him with a unique showcase. Bell developed an iron-box receiver in spring 1876. It was this device, used with both variable-resistance and electromagnetic transmitters, that he set up in late June in the exhibition's main building to demonstrate to a group of prominent American scientists and—important for publicity—Brazilian emperor Dom Pedro. Bell's voice sounding over the device amazed both experts and emperor alike. The success made Bell's telephone the talk of the nation's scientific community, and visiting foreign scientists carried the word abroad. Bell then undertook a series of lecture-demonstrations around the Boston area and in Brantford with improved equipment. Local newspapers soon awoke to the discovery, and Bell became a popular and effective advocate for his brainchild.

At first hearing, the telephone was a memorable and even frightening experience for people. Some rural audiences almost panicked when they heard a disembodied voice coming from the iron box. But people quickly accepted it. It worked. The telephone quickly became commonplace. The first regular telephone line was established on April 4, 1877, from the shop where telephone equipment was manufactured to the suburban Boston home of the shop owner. In July when the Bell Telephone Company was formed, over a hundred telephones were in operation, and by August the number jumped to six hundred. (At Bell's death in 1922, the company served well over nine million telephones.)

Yet Bell was not interested in running a business and soon withdrew from the company's board of directors. He gradually ceased to participate in company affairs other than serving as a witness in various patent litigations. By 1881, he had severed his connections with the company, only retaining some two thousand shares of stock. Bell had previously sold the bulk of the family stock during the boom market for telephone shares. The growth of the business, eventually reorganized into the American Telephone and Telegraph Company, was the result of other men's efforts. Nonetheless, it was Bell who was called on to inaugurate telephone service between New York and Chicago in 1897—an event captured in one of the best known of Bell photographs.

Bell's telephone over the years became an integral part of the American household. Popular legend has it that Bell himself refused to have one in his own home, and endowed with a rather-playful sense of humor, Bell at times encouraged the story. As a matter of fact, he did have telephones in his house but not one in his study. Perhaps the most important immediate effect of the telephone, however, was to facilitate the burst of urbanization that took place in the 1880s. The rapid growth of modern cities and industry depended on the telephone for instant communication without the delays as well as congestion of hand-carried messages.

The telephone brought Bell fame and fortune before he was thirty. He never ceased inventing, but in his later discoveries he was never able to

For the opening of the Columbian Exhibition on October 18, 1892, Bell inaugurates the first New York to Chicago line.
Source: American Telephone and Telegraph Company

equal his first achievement. Bell brought together an unusual complex of knowledge, ability, and character in his pursuit of the telephone: speech, music, acoustics, mechanics, and electricity all mixed in his mind, and combined with ambition, drive, and an eloquent, compelling personality. His inner need to achieve was strong, and he had pushed himself to make the telephone work in order to marry Mabel Hubbard. With marriage and success, Bell's interests and drive were free to follow many paths. Bell next conceived of sending the voice by a light beam and developed a wireless "photophone." He used the element selenium, which varies in its ability to conduct electricity depending on the amount of light falling on it. A transmitter with a voice-vibrated mirror produced a beam of light with a fluctuating intensity. The beam was focused on a selenium receiver, which in turn varied the current to a telephone receiver. Bell succeeded in transmitting sound by sunlight but only so far as the beam would go unbroken. The photophone was not practical without a technological breakthrough like today's laser. But Bell's photophone work opened up the way for the nondestructive analysis of opaque substances by means of light rays: photoacoustic spectroscopy—a technique that has only been recently utilized.

As an aid to surgery, Bell developed a telephonic probe that located bullets or other metal in a body. The probe was successfully used until the X-ray supplanted it during World War I. Also in the field of medicine, Bell invented the "vacuum jacket," a mechanical device placed around the chest that administered artificial respiration. The concept was later revived in the modern iron lung.

After winning France's prestigious Volta Prize in 1880, Bell organized the Volta Laboratory with two other engineers and inventors. They turned to the phonograph, which Edison had abandoned to begin his electric light experiments. The Volta associates developed a new method to record sound by cutting grooves in waxed cardboard cylinders as well as a floating (rather than indenting) stylus to reproduce the recording. They successfully marketed the patents, and Bell used his share to establish a trust fund for research on deafness. Yet it was there that the group's collaboration ended.

His imagination captured by mechanical flight, Bell joined a physicist friend, Samuel Langley, in his efforts to build a flying machine during the 1890s. Bell also experimented with lighter-than-air craft and kites to work out aerodynamic and structural problems. In 1907, inspired by the Wright Brothers' flights, Bell gathered a group of young engineers, including future airplane manufacturer Glenn Curtiss, to form the Aerial Experiment Association. In designing their airplane, they devised a means to achieve lateral control using "ailerons" or hinged wing tips that moved in opposite directions. They were later to discover that similar mechanisms had already been developed in France. But the group had only agreed to associate for a year, and although the men extended their work for six months to achieve a flight of some three and a half miles at fifty feet, they disbanded in 1909.

While working with kite structures, Bell recognized the possibilities of the tetrahedron, a geometric shape of four equilateral triangles forming a pyramid with a base. He envisioned a new kind of architecture, combining both strength and lightness. The advantage of the tetrahedron is that its sides, as triangles, are all compression or tension, and cannot be warped. Structures made up of tetrahedrons are braced in three dimensions. The individual members can be mass-produced, such as stamped out of metal, and then bolted together. Tetrahedral construction, moreover, requires relatively little money, time, and skill, yet it can construct great spans for roofs or bridges. Bell's concept foreshadowed Buckminster Fuller's geodesic domes, and today it has come into wide use as space frame architecture. Bell patented tetrahedral construction in 1904 and built a tower at his Canadian summer home to demonstrate the potential. But the idea did not take, and Bell did not pursue it.

Another offshoot of his aerodynamics research was Bell's interest in hydrofoil boats. With fellow Aerial Experiment Association member Casey Baldwin, he built a series of "hydrodromes," the last of which set a world marine speed record in 1919 at 70.86 miles per hour. Yet neither the U.S. nor the British Navy was interested, and the Bell-Baldwin Hydrodrome Company failed to find any business. The hydrofoil patents granted in 1922 were to be Bell's last.

Bell anticipated today's technology in many ways with these inventions, but his later ideas have a sense of anticlimax when compared to the telephone. Bell's interests were scarcely limited to invention, however, and it was in these varied pursuits that he found meaning.

If Bell's life was particularly concerned with communication and breaking down the barriers between people, this commitment was nowhere more evident than in his concern for the deaf. In his own family, both his mother and his wife, Mabel, were deaf. Mabel had lost her hearing at the age of five and had come to Bell as a pupil in 1873. They were married four years later. An accomplished speech reader, she had learned to articulate relatively well, if not perfectly. She often astonished new acquaintances with her ability to follow their conversation.

As a teacher of the deaf, Bell's ideas embroiled him in a lifelong controversy that still continues among those who work with the deaf: the extent to which sign language as distinct from lip or speech reading should be used. (The form of sign language most widely used in Bell's day had been devised by the Abbé de l'Épée in the eighteenth century to express ideas rather than just finger spelling.) In general, sign language is less prone to misunderstanding—that is, it does not require the context to fill in the meaning. It is easier, especially for those who are congenitally deaf, and perhaps more natural. Speech reading is much more difficult, relying heavily on quickness of mind to grasp the context and—most important—requiring the user to be familiar with the language being spoken. Bell's argument against sign language was that it was a specialized language that cut the deaf off from communication with everyone except those who could use it.

He also felt that it deprived the deaf of a full understanding of English—or other languages—since sign language falls far short of English in conveying abstract ideas and has a limited subtlety of expression.

As a corollary to speech reading, Bell emphasized teaching the deaf to articulate, think, and communicate in English. His beliefs brought him into frequent conflict with Edward Gallaudet, noted educator of the deaf and originator of the "Combined Method," which made extensive use of sign language although employing other methods as well. Bell's commitment also led him to help found and endow the American Association for the Promotion of the Teaching of Speech to the Deaf in 1890. Renamed the Alexander Graham Bell Association for the Deaf, it still serves as a world-wide information center and library, administering grants and publishing research relating to deafness.

Another of Bell's varied legacies is the *National Geographic Magazine*. Out of a sense of family loyalty, Bell assumed the presidency of the National Geographic Society in 1897 on the death of Hubbard, who had founded the organization. To stimulate a declining membership as well as revitalize the society's dry and pedantic magazine, Bell found an able, young editor in Gilbert Grosvenor. Bell's financial contributions and ideas for eye-catching, dynamic illustrations provided the basis for Grosvenor's sweeping innovations in the magazine's format. The new editor created the successful mix of photographs and factual, easy-to-read articles that is the hallmark of the magazine even today. He later married Bell's elder daughter, Elsie, and the family is still active in the magazine.

Bell's wide-ranging interests, active participation in many causes and organizations, and gift for public speaking gave the impression of an expansive personality. People constantly sought him out. He seemed almost larger than life with a tall, imposing figure, resonant voice, dark eyes, dark complexion, and full beard—its black growing white as he aged. Yet all his life he struggled against a tendency to withdraw. His chosen hours of work, from late at night through to early morning, reinforced his solitude. He felt a natural sympathy with the isolation of the deaf. More at home in the little world he created for himself on his Nova Scotia summer estate, Bell nonetheless led an active social life in Washington, D.C., where he had settled in 1880. Mabel did much to counter his aloofness, as did his own awareness of his reserved disposition.

Bell sought communication for the world at large, the deaf in particular, and himself as an individual. To Americans his name—so appropriate—is inextricably linked with the telephone. Indeed, at the time of his burial, all telephone service in the United States was stopped for a minute in tribute. Certainly Bell's achievements place him at the center of the communications revolution that so immediately links people at opposite ends of the globe together.

11 Thomas Alva Edison and the Rise of Electricity

Thomas Parke Hughes

At the same time that Thomas Alva Edison flourished, the United States emerged on the world scene as the great technological nation. This simultaneity was not altogether accidental, for Edison at once drew on the sustaining environment and helped create it. His most fruitful years were those spent in Menlo Park, New Jersey, from 1876 to 1886, which was about the time America rose to preeminence in invention and industry. By 1890, the United States led the world in the number of patents granted as well as its iron and steel production. Furthermore, its production of coal—the basic fuel and an important chemical—ranked second to none. Of the many inventive Americans in this productive era, Edison was the most prolific with no less than five hundred patents by 1885, and hundreds more to follow. Not only did he have the largest number of patents, but the devices and processes they covered were financially rewarding and technologically impressive.

No wonder that Edison was the American hero of the Gilded Age. Newspaper coverage showed him to be one of the most interesting of men to Americans, and a popular poll conducted during his lifetime revealed him as the man that Americans most admired. Americans of his generation liked and admired him because he had risen from humble circumstances to a position of affluence and fame, and because he provided them the material needs and pleasures for which many had come to the New World.

Born in 1847 in Milan, Ohio, a midwestern town earning its living from a canal along with the processing and shipping of agricultural products, Edison was the son of a small manufacturer of wood shingles and a schoolteacher. Later the family moved to Port Huron, Michigan, and there, at age twelve, he became a traveling candy and newspaper "butcher" (salesman) on the railway to Detroit, setting up his small laboratory for science experiments in the baggage car. This episode and other activities of young Edison's life became national folklore after his world fame provoked a host of biographies. Parents urged their children to emulate him, and through him, these parents lived vicariously the American dream.

The anecdotes, once so well known, now recede into the background of the past as a worldly America seeks other representative types and even antiheroes. So we should briefly recall that the laboratory on the car to Detroit had a fire that compelled young Edison to experiment elsewhere. Also we might remember that he was introduced to the telegraph by a grateful operator whose small son Edison had pushed from the path of an oncoming train. And that his partial deafness began to isolate the young man from his immediate environment and turned him toward introspection. To understand him, we should note his Wanderjahr as a journeyman telegraph operator, for he learned not only the art of survival in a tumultuous and demanding world but also a great deal about electricity. He was thought a bit odd by his fellow operators for choosing to stage experiments according to the precepts of the great Michael Faraday, whom Edison much admired and deeply read, instead of joining his fellows in a night on the town, wherever that might have been for the highly mobile telegraphers.

By 1868 he was working for Western Union in Boston, and there he committed himself to invention. At first his inventing could only be during off-hours, but he found time to build and patent an automatic vote recorder for which he could find no market. (He later said that it was then he decided to identify a market always before he invented—an obvious strategy employed by most professional inventors, then and now.) No amateur, and determined to live by his new profession, Edison journeyed to New York in 1869. With him he carried ideas for improvements in telegraph systems, and had a sharp eye and clear head for opportunity. Opportunities opened to him shortly when chance allowed him to repair a Wall Street printing telegraph at a time when its price quotations were badly needed. Subsequent support from the grateful owner, who was well connected in the telegraph business, gave Edison an entrée. Less familiar are accounts of the intricate business and technological activities in which he became involved as he invented telegraph improvements, including a quadruplex design, for several, even competing, telegraph companies. When Western Union fell into the hands of Jay Gould, the notorious financier and waterer of stock, Edison said his kind of inventiveness was no longer needed there, so he became an independent inventor, choosing his own problems, making his own inventions, and forming new companies to market them. In 1870, he had set up a telegraph manufacturing shop and laboratory in Newark, New Jersey; in 1876, when he decided to become an independent inventor, he drew on his capital and growing reputation to fulfill a vision: the establishment of an invention enterprise or, as some said, invention factory.

He chose Menlo Park, a lonely site on the Pennsylvania Railroad between New York City and Philadelphia, which as bases of supply and ready markets were only an hour or so away by train. But in Menlo Park, unlike the cities, there was a freedom from worldly distraction and invitation to concentration. Edison realized this as he moved old experienced aides such as John Kreusi, the machinist and ingenious model maker, from the Newark shop, and brought in new ones who would have to learn the Edison style and absorb the deep commitment to inventing a method of invention.

The compound in Menlo Park was both cozy and workmanlike. The buildings provided the resources needed by a professional inventor, and Edison soon became known as the Wizard of Menlo Park. Within a few years of settling at the new site, Edison had a building for an office along with a technical and scientific library (long series of the world's leading journals were housed there for the seekers of ideas about the state of the art) as well as another large building that housed on its two floors a remarkably well-provided chemical laboratory, electric testing facility, and initially, machine shop. (Later, the machine shop—the producer of electric and mechanical models, small and full scale—was separately housed.) After Edison concentrated on the invention of a system of electric light, small buildings for blowing the glass bulbs and obtaining the filament carbon were added. A carpentry shop rounded out the facilities. When the invention factory was built and full of life, a number of watercolors and drawings captured the public imagination by portraying it snow covered, suggestive of a bountiful Santa Claus and his busy elves. In fact, the place worked more like a center for advanced invention.

The sociology of the Menlo Park group deserves more study. It was far more complex in its interactions than the popular impression of the inventive genius delegating work to eager, pliable assistants. Thousands of laboratory notebooks from Menlo Park suggest that Edison, sensitive to innumerable factors, decided on the ultimate objectives, but that a handful

Thomas Alva Edison in the laboratory in Menlo Park, New Jersey, exhibits the forerunners of modern radio tubes developed by him and his team of researchers.

of men immediately around him engaged in team research and development, coordinated and monitored by him. The historical record also reveals
that the members of the inner circle occasionally changed, as did the equipment at Menlo Park, in accord with the nature of the project.

From 1878 to 1882, when activity concentrated on the invention and
development of a system of electric lighting, Francis Upton, Kreusi, and
Charles Batchelor appear frequently in the notebooks carrying out experiments and recording observations. Their particular characters and work tell
us much about Edison's. Upton's presence discounts the belief that Edison
did not value science. A physicist, Upton had studied at Bowdoin College,

Princeton University, and in Berlin under the famous Herman von Helmholtz, before coming to Menlo Park at the start of the electric light project to read about the state of the art in the foreign scientific and technical literature. Soon he had charge of the development of the dynamo for the system. Batchelor, another intimate, was so deeply involved that some said Edison suspended work when Batchelor absented himself from the laboratory. Batchelor was an Englishman and expert mechanic who had first come to America to work on textile machines. Kreusi, also a machinist by trade and European by origin, was the master model builder—a function without which Edison and other inventors could not work. It was said that Edison had merely to tell Kreusi of a device to have him reduce it to drawings and a model. (Kreusi is best known for his rendition of the first phonograph.) Edison, similar to other professionals in technology and industry, drew on British and European technology and science as he surrounded himself with men who had apprenticed or studied in the Old World.

Edison was the leader of an invention and development group (today it would be labeled research and development), but he was also something more: an inventor-entrepreneur. Today, only a few men attempt to carry such a broad range of responsibilities as designated by that term. Edison not only presided over invention and development but he also took part in financing, publicizing, and marketing for the project. His most famous project, the electric light system, serves well to illustrate the point.

To choose a problem or project is a critical decision for an independent inventor; an inventor hired by a corporation usually has guidelines explicitly defined or implicitly revealed by the vested interests of the corporation or agency for which he works. Edison decided, for complex reasons, to expend his resources on the electric light project in 1878. Friends in science and engineering told him that the state of the art in incandescent lighting suggested that practical achievement might be near. Technical periodicals and patents also signaled activity in incandescent lighting. Such information alerted Edison to the possibility that he might solve the remaining critical problems—such as a durable filament—that would make the difference between ingenious tinkering and commercial success. He had confidence in his ability to solve electric lighting problems because like so many professional inventors, he knew his characteristics and drew on the experiences that had helped shape them. In short, he was, after years of work on the telegraph, an expert on electric matters. The electromagnetic phenomena of the telegraph, electrochemistry of the battery, fine mechanics of the relay, and laws of circuitry all had relevance to the new endeavor, if one could transfer, adapt, and invent by analogy.

Another reason for working on electric light was the nature of Menlo Park itself, both its physical and personnel resources. For its day, Menlo Park represented a considerable investment of resources, and as a result, it had substantial momentum. It had mass, movement, and direction. Therefore, certain problems could be solved best at Menlo Park, and others better elsewhere. Edison and his advisers realized that the problem of inventing and developing an electric lighting system suited Menlo Park, for a system involved electromagnetic machines (generators), delicate apparatus (switches, fixtures, controls, incandescent lamps, and so on), and complex circuitry. Because of the diversity of problems posed by the varied components, the complex of facilities and people at Menlo Park could be advantageously employed. The system required the repeated testing and experimentation for which Menlo Park and its men were also well suited. Most important, a system of electric lighting needed vision, planning, and coordination, for which Edison had a genius and Menlo Park was designed.

Before finally committing himself, he had to define the market and identify the financial resources. Edison, who often thought in analogies, saw that an incandescent lighting system was like a gas one, and he knew that gas lighting thrived commercially. Therefore he would be following the less risky course of improving, instead of introducing, a product or service. To make sure, he commissioned a survey of gas lighting use in the thickly populated Wall Street district of New York City. There the deep, man-made canyons between the office buildings and thousands of offices within needed artificial light. It was not by coincidence that he focused on offices owned by men such as J. P. Morgan, who had the funds to finance Edison's project—and did.

The project needed a business and financial structure as well as the technological one provided by Menlo Park. So with the advice of Grosvenor P. Lowrey, whose strong characteristics as an experienced business and financial entrepreneur nicely rounded out his own, Edison established the Edison Electric Light Company in fall 1878. The company's purpose was to fund the inventive enterprise of Edison insofar as it pertained to electric light and power, and promote the adoption of the patented inventions

throughout the world. There is much to be learned about the art of technological innovation from the constitution of this company. Edison's investment, for which he received stock, was his form of capital—his patents. He
assigned to the company for five years all his patents in the general category of electric lighting and power. Morgan and other financiers contributed their resources—cash. Edison awarded his most valued staff members,
such as Upton, stock in the other companies that were soon established.
These companies manufactured components and performed services that
would be needed in the Edison system of electric lighting. (Unlike some
organizers, Edison did not impose an administrative structure on the creative concept.) The companies institutionalized the various components
or functions of the system: the Edison Electric Illuminating Company of
New York (1880), founded to preside over the first demonstration central
station; the Edison Machine Works (1881) to manufacture the generators;
the Edison Lamp Works (1880); the (Edison) Electric Tube Company (1881)
to make the underground conductors for the distribution systems; and
Bergmann and Company to produce electric light accessories. Other inventors who came into the lighting field were frequently willing to modify the
designs of existing generators, or turn over manufacture of components to
companies not under their direction and control.

The history of the invention of the Edison electric lighting system is
well known and need not be repeated in detail here. It should be pointed
out, however, that far too much emphasis in most accounts has been
placed on the search for—and testing of—the lamp filament. This stress
distracts attention from the essence of Edison's genius. He wrote of his
approach:

It was not only necessary that the lamps should give light and the dynamos generate current, but the lamps must be adapted to the current of the dynamos, and the
dynamos must be constructed to give the character of current required by the lamps,
and likewise all parts of the system must be constructed with reference to all other
parts, since, in one sense, all the parts form one machine, and the connections between the parts being electrical instead of mechanical. Like any other machine the
failure of one part to cooperate properly with the other part disorganizes the whole
and renders it inoperative for the purpose intended.

The problem then that I undertook to solve was stated generally, the production
of the multifarious apparatus, methods and devices, each adapted for use with every
other, and all forming a comprehensive system.

Rarely does one find so succinctly stated a systematic approach to invention.

By September 1882, the system had been conceived, designed, patented, and tested on a small scale at Menlo Park. Various Edison companies manufactured the equipment. The first Edison central station for public supply was then placed into operation, serving, as planned, New York's Wall Street district with about a one-mile radius. The Pearl Street Station had six steam engines, driving six Edison Jumbo generators, each capable of supplying twelve hundred 16-candlepower lamps. Within one year, about eight thousand Edison lamps were being supplied from a 110-volt distribution system. The world celebrated its first central station's technical success.

Financial reports, though, show that for the first few years, the Pearl Street Station sold electricity at a loss. This was sustained for several reasons. Foremost was the consideration that Pearl Street served as a demonstration plant to interest local civic leaders and financiers throughout the country as well as abroad in buying the licenses to operate an exclusive franchise along with the equipment of a central station similar to Pearl Street. Another reason was the valid assumption that as service was improved, customers were added, unit fixed costs were lowered, and various economies were achieved through rationalization, the operation would then become profitable. It did, before fire destroyed the historic station in January 1890. By then, there were Edison stations in large cities and small towns throughout the world. The era of the Edison direct-current station had been established; this period in the 1890s gave way to that of the alternating or polyphase station serving a larger area with both power and light over high-voltage systems. Edison did not demonstrate the flexibility to make the transition.

After the triumph at Pearl Street, Edison's period of brilliance passed. He lived and worked on until 1931, adding to this long list of inventions and patents, and making substantial innovations, but his later contributions to the technology of motion pictures, magnetic ore separation, portland cement manufacture, the storage battery, and the derivation of rubber from indigenous American plants lacked the incisive insight as well as dramatic rendition of his work on the quadruplex telegraph, telephone transmitter, and early phonograph, all of which—like the lighting system—came before 1882. Perhaps a key to the apparent diminution of the inventive powers can be seen in the considerable and unsuccessful effort he made to introduce a process of magnetic ore separation.

Following Pearl Street, other events in Edison's life added up to a watershed—the gradual turn downward after the peaks of achievement. In 1884, his first wife died of scarlet fever, and some of his friends believed that Edison, deeply grieved, then lost his taste for Menlo Park. In 1886, he moved into a new laboratory of his own design in West Orange, New Jersey. It was much larger and more complex than the rural compound so easily pervaded by his personality. Also in 1886 he married Mina Miller, an attractive young socialite from Akron, Ohio, for whom he purchased an estate, Glenmount, in the hills above his new laboratory. (His first wife, Mary, had worked in his Newark shop, and they had lived in a relatively simple house in Menlo Park.) Between 1882 and 1892, he was also losing influence in his electric manufacturing companies, and at about the time he embarked on the ore project, he finally sold out his stock in the enterprises and saw them consolidated in 1892 not as Edison General Electric but instead as simply General Electric.

Income from the sale of the electric stock financed the magnetic ore separation project. Edison decided by a market analysis, reminiscent in its thoroughness of that made in the Wall Street district for the Pearl Street Station, that there would be a return on the investment. He concluded that if a low-cost process for concentrating the low-grade magnetite ores of the Northeast could be developed, then because of lower transportation costs, these could displace in the eastern steel mills the naturally richer ores of the Upper Peninsula of Michigan. He calculated that the concentrated ore could be sold competitively and profitably at the eastern mills for $6.00 to $6.50 per ton. In order to minimize the cost of concentration, he decided to exploit the economies of scale. Edison invented and developed a massive system of magnetic ore separation. In essence the system broke down large aggregates of rock, sand, and ore into fine particles, and then passed them between the poles of a series of powerful electromagnets that drew off the highly magnetic ores and thereby concentrated them. There were innumerable technological problems to be solved, but the process envisioned at the cost projected was in operation after five years in 1899. The catastrophe for Edison was that by then, the fabulously rich ores of the Mesabi range in Minnesota were being exploited, and these could be delivered to the eastern mills below his costs. He shortly abandoned the project, having lost at least $2,000,000.

Edison admirers insisted that he had succeeded by technological standards because the process worked as intended. But Edison did not judge himself by these standards; he was an inventor-entrepreneur, and as such, viewed finance, business, invention, and engineering as inseparably intertwined. Why, then, had he failed as an inventor-entrepreneur? The most obvious answer is that his market projections had not been accurate. Other reasons, however, suggest themselves. In embarking on the ore separation process, he abandoned familiar ground. Earlier he chose problems demanding brilliant conceptions of elegant devices and machines, such as telegraph instruments along with the telephone apparatus, electric lamp, dynamo, and phonograph. He had not proven himself remarkable as an inventor of large-scale manufacturing systems. Furthermore, there was an excitement about his earlier inventions shared by those who used them and the man who invented them. The design and organization of the ore separation process was a grim and grimy enterprise. For almost five years, he and hundreds of workmen labored in the bleak highlands on the New Jersey–Pennsylvania border. One can only ask how much of the Edison spirit and enthusiasm, so clearly displayed in the Menlo Park notebooks, worked for him on the ore project. Also missing were the fine craftsmen and electricians who worked so near him and so well at Menlo Park. No Kreusis, Batchelors, or Uptons emerged as salient figures from the ore separation project. In addition, Edison was losing his way in entering a field where brute size—technological, financial, and business—counted more than ingenuity and finesse. Elmer Sperry, another noted inventor and contemporary of Edison's, remarked that he always chose the hard and fine problems in order to avoid vulgar competition.

Despite the outcome of the ore separation project and lack of brilliance surrounding later projects such as the storage battery, Edison rode the crest of public acclaim until by the time of his death, he approached the status of a secular saint, a representative to masses of Americans of the best in the American character. They saw him as a plainspoken, self-educated, practical minded, eminently successful, native genius. They believed him to be an inspired empiricist, both an experimenter and tinkerer. He was an American success story—that of a hardworking, hard-nosed, down-to-earth man who provided material in abundance for an upwardly mobile society. The public wanted more of him than inventions; his pronouncements about education, religion, and other general questions commanded front-page newspaper space, and became oracular statements for countless admirers.

In fact, Edison was more complex. He spoke plainly when raising money from Wall Street financiers who distrusted long-haired scientists; he was self-educated, but his reading included the classics of Western literature and Faraday's notebooks; he was eminently successful, but not simply because he had inventive genius; and truly, he practiced the art of the experimenter with consummate skill, yet he also knew the science of his day, and used it to formulate hypotheses and organize experimental data. In the laboratory with his associates, he could be described as "the wizard that spat on the floor," but it is unlikely that the floors were so stained at Glenmount. Long before professional public relations perfected image making, Edison presented to the world an inventor whom it would support and hero whom it needed.

During his lifetime, the center of industrial research moved from Menlo Park and West Orange into the General Electric, Bell Telephone, and DuPont Laboratories. Men with advanced degrees in science working in the laboratories did not merely use the available science to solve their technological problems; they generated the science as needed. Americans, by the time of Edison's death, began to look to men in white coats bending over microscopes as images of the new research and development bringing "better things for better living."

Despite his limitations, however, Edison was nonetheless a representative American and key to understanding its late nineteenth-century character, when it became the world's leading technological and industrial power. America, then, had moved beyond steamboats, railroads, and textile mills; it had not yet reached the stage of automatic controls, missiles, and computers. Edison possessed just the touch to provide the lights and sounds that brought a sense of well-being and even affluence to hardworking people still laboring on the construction site that was young America.

12 Lewis Latimer and the Role of Black Inventors

Carroll Pursell

Lewis Howard Latimer was one of the most successful African American inventors at the turn of the twentieth century. His life (1848–1928) encompassed the last years of slavery, Civil War, promise of Emancipation, hope of Reconstruction and its brutal betrayal, and crushing imposition of an era of segregation. It also covered the years when industrial capitalism triumphed in America, and a second industrial revolution saw the dramatic birth of such new technologies as electric production, distribution, and widespread adoption. Latimer not only lived through all of this but his life was decisively shaped by these events as well: he was a black man who carved out a career in the growing (white) electric industry, first as draftsman, then as inventor, and finally as manager and expert witness.

The subject of black inventors has been contentious throughout American history. On the one hand, the denial of inventive abilities in enslaved Africans had been an important prop to racist limitations imposed on them. To counter this canard, it had long been seen as crucial that evidence of black invention be discovered and celebrated. The resulting record of black invention has tended to be recuperative, liberal, and individualist: recuperative in that it has sought out forgotten and neglected inventors (like Latimer), liberal in that it assumes the progressive nature of both invention and inventors, and individualist in that after a brief acknowledgment of the historical, structural, and cultural handicaps under which such people labored, their stories are told as narratives of individual triumph over adversity. All three tendencies have worked to cast black inventors as "heroes" as well as role models for the young.

Attempts to document and thereby celebrate African America inventive activity went back at least as far as 1834, when the editor of the pioneering black newspaper the *Liberator* called on "colored inventors" to send him information on their work. He wanted, he said, "to collect proofs of colored talent and ingenuity in the United States." Eight decades later Henry E. Baker, in his book on *The Colored Inventor: A Record of Fifty Years*, claimed that some twelve hundred "instances" of patents had been collected, although he was only able to verify that eight hundred had been the work of black inventors. These eight hundred, he wrote, "tell a wonderful story of the progress of the race in the mastery of the science of mechanics."

It is important to realize, however, that despite the special roadblocks to black participation in technological innovation, invention as an activity of both black and white participants shared significant common characteristics. From the late nineteenth century to the present, technology in America has become increasingly science based, making ever more difficult the career of the practical tinkerer, and this has been especially true in the electric industry. The journal *Scientific American* editorialized in 1894 that "those who pursue electricity, by the inductive method—that of practical

Lewis H. Latimer.
Source: The Queens Borough Public Library, Archives, Lewis H. Latimer Papers

experiment and generalization therefrom—have lately entered the complaint that electricity has become nine-tenths mathematics. To many real experimental discoverers, the higher branches of mathematics constitute a trackless maze." At the same time, and partly as a result of this trend, the rising cost of research and development has meant that inventions have increasingly come from corporate laboratories rather than the garage workshops of popular fame. And finally, as has always been true, most inventions do not result in any significant innovation in practice and simply slide into deserved obscurity. Latimer's technical career was shaped by all these factors.

Latimer was born in Chelsea, Massachusetts, in 1848, the son of George Latimer, who had escaped from slavery in Virginia in 1842, along with his new wife, Rebecca, who was also enslaved. Soon after the couple reached Massachusetts, George's former master arrived to claim him and force his return to Virginia. George's cause, though, was taken up by a group of prominent abolitionists who publicly denounced the injustice and successfully raised the money to pay off his pursuer. The father's resulting fame and his celebration within the abolition movement probably had a role in shaping his son's later view of the possibilities of benign interracial relations.

Latimer received both a primary and secondary education, but began work when he was thirteen, including a stint as an office boy for a prominent local attorney. In 1864, at the age of sixteen, he enlisted in the Union Navy and was discharged the following year. After taking various jobs, he answered an ad for "a colored boy with a taste for drawing" placed by the firm of Crosby, Halsted, and Gould, "Solicitors of American and Foreign Patents." While working as an office boy, Latimer carefully observed the work of the draftsman who made the patent drawings for clients, and bought himself a book on drafting and then a complete set of drafting instruments. After "looking over the draftsman's shoulder to see how he used his instruments," he later wrote, he would practice at home. When he felt sufficiently confident of his abilities, he asked the draftsman to let him make some of the drawings, and his work was good enough that when the draftsman resigned his job, Latimer was appointed as his replacement. Not only did he make drawings as part of his new post but also supervised the construction of the models then required by the Patent Office.

Over the eleven years that Latimer worked for the firm, all the time gaining experience, knowledge, and responsibility, he also began to try his own hand at invention. His first patent (No. 147,363), taken out in 1874 with a C. W. Brown, was for "Water-Closets for Railroad Cars." Around this same time, he did some of the drafting (either as part of his job or on a freelance basis) for the telephone patent application of Alexander Graham Bell. Over the period 1874 to 1905, Latimer was granted eight patents, and for an uncounted number of other inventions he did not, for one reason or another, apply for patents.

After a management change at renamed Crosby and Gregory, Latimer left the firm in 1878. He briefly worked with another Boston patent solicitor, but then bad economic times saw him trying to piece together a living working as a paper hanger, painter, and briefly in an iron foundry. At his sister's suggestion he moved to Bridgeport, Connecticut, a city he later described as "perfectly alive with inventors" and "a perfect hornet's nest of industries." Here he took work again as a draftsman, this time in a machine shop. As luck would have it, he was seen one day by Hiram Maxim, the celebrated inventor of, among other things, the Maxim machine gun. Indeed, Maxim had patents for such items as a mousetrap and steam pump and worked at trying to develop an airplane for powered, heavier-than-air flight. Among Maxim's many enthusiasms was the nascent field of electricity, and in 1880, he hired Latimer as a draftsman and personal assistant in his U.S. Electric Lighting Company. The new job placed Latimer firmly on the path that was to lead to his eventual prominence in the industry.

Latimer had begun his career as a draftsman, and had invented and patented a number of devices, but spent the last years in the industry performing services that might have been more reasonably expected from an electrical engineer. Part scientists and part businesspeople, engineers often professed a disdain for "mere" theory and "book learning," preferring to privilege practical and accumulated experience. The definitions of that profession were still fairly fluid, and qualifications could be variously acquired, but it was not one welcoming to anyone who was not white, male, and of at least the respectable working class. In fact, one engineer warned in 1903 that "we have the man who fires the boiler and pulls the throttle dubbed a locomotive or stationary engineer; we have the woman who fires the stove and cooks the dinner dubbed the domestic engineer, and it will not be long before the barefooted African, who pounds the mud into the brick molds, will be calling himself a ceramic engineer."

This fear that the presence of women, workers, and people of color would lower the status of engineering was structured into the way engineers were educated and organized. The American Institute of Electrical Engineers, for example, which had been organized in 1884 (the year before Latimer went to work for the Edison Electric Light Company), appears not to have had an application for membership from a black engineer until 1903. When such an application was finally received, the institution to its credit decided that it should "take the regular course," and went on to specifically stipulate that the "Board of Examiners should consider all applications solely on their technical merits without reference to color or sex restrictions." The next year, however, when an application was received from a female engineer, it was set aside while a special category of women members was suggested. The application was not reconsidered.

Similarly, formal education in engineering was well established by the time Latimer began his career. Rensselaer Polytechnic Institute was the world's first private engineering school in 1829, and Harvard set up a separate engineering school in 1842, as did Yale in 1847. The land grant universities set up by states after 1862 all had significant engineering schools, and American engineering schools numbered 100 in 1893, and then grew to 160 by 1933. The Massachusetts Institute of Technology, perhaps the nation's best-known engineering school, was established in 1861. The first-known black student to attend MIT entered with the class of 1888 and graduated in 1892. But in fact, engineering was, and remains, a bastion of white male privilege; in 1986, only 1.7 percent of the country's scientists and engineers were African American, and in 1989, only 17 percent of engineering degrees were awarded to women. When Latimer went to work for Maxim in 1880, however, most engineers were neither university trained nor members of any professional organization.

Maxim's company, then only in its second year, concentrated on manufacturing and attempting to improve his incandescent light bulb, which came on the market only a year after Edison's. The company, including Latimer and his family, moved to New York City in 1880, and that same year made its first commercial incandescent lighting installation. Latimer was kept busy not only drafting but also working at the lamp manufacturing effort and helping supervise installation of the Maxim systems. The company was feeling its way along in part because unlike Edison, who had hired a university-trained mathematician, it had no formula for calculating the electric load that various wires could safely carry, and the whole decision on this critical matter had to be determined by experience and a sense of the unique situation.

Patent No. 247,097 for an electric lamp, issued Sept. 13, 1881 to J. V. Nicholas and L. H. Latimer

The more Latimer did this sort of work, the better he became at making useful decisions. It was a small company, and Latimer was called to move from one task to the next, becoming a jack-of-all-trades. He was sent to Philadelphia and Montreal, for example, to supervise new installations, was placed in charge of producing carbons for lamp filaments, and even found time to continue making and patenting improvements in lamp manufacture and design. In 1881, he was sent to England to help set up a lamp factory for the new Maxim-Weston Electric Lighting Company, where he saw to the training of the workmen and smooth operation of the plant. The facility was successfully productive within nine months, and the Latimers returned to the United States.

On his return, Latimer faced another of his periodic bouts of job hunting and underemployment. He was both advantaged and defeated by the large number of small, usually vulnerable companies that had sprung up in this rapidly expanding as well as evolving industry. A short stint with the Weston plant in New Jersey was followed by a better job as draftsman and manager of lamp production for the Olmstead Electrical Light and Power Company in Brooklyn. Here he continued his improvement in incandescent lamps, not all of which were patented or even particularly improvements. When the Olmstead firm failed in 1883, Latimer moved to the Acme Electric Lighting Company, which in its turn, failed in 1884. He next went to the Excelsior Electric Company, but within a few months was hired by the Imperial Electric Light Company. In 1885, he was briefly with the Mather Electric Light Company, but then before the end of the year, was hired by the Edison Electric Light Company, which in its program of innovation and economic success, stood in marked contrast to all Latimer's previous employers.

Edison was beginning a period of aggressive legal action against those companies that it believed were infringing its patents, especially that iconic 1880 patent for Edison's incandescent lamp. Latimer had held responsible positions, usually having to do with the improvement and manufactures of lamps, with a number of Edison's rivals, and was soon assigned to the engineering department where he was to be a "draughtsman inspector and expert witness as to facts in the early stages of the electric lighting business … [to travel] extensively, securing witnesses' affidavits, and early apparatus, and also testifying in a number of the basic patent cases to the advantage of his employers." It was a job demanding deep and broad knowledge of the technology and businesses involved along with a keen sense of how best to conduct oneself and deploy that knowledge. That an African American could be picked for and succeed at such a specialized as well as sensitive task during a period of rapidly deteriorating race relations in the country seems astonishing.

Latimer was moved to Edison's legal department in 1889, the year in which the company sued the U.S. Electric Lighting Company for patent infringement. He had worked for this company some years before, and it was using patents from both Maxim and Weston in its lamps. Edison won its case in 1891 and then the appeal, which was rejected a year later. Latimer's expertise would certainly have been useful in this case and others that followed. In 1896, he was also made the chief draftsman of the Edison Electric/Westinghouse Board of Patent Control, a body set up by the two giants in the field to handle cross-licensing agreements between them. It was his last appointment by the Edison Company.

Latimer had been thirty-seven when he took his new job, only a year younger than Edison himself. His role at Edison was clearly to help the company shape the past rather than the future of the industry and concern himself with "early apparatus" instead of new innovations. He did continue to invent, and despite having to take account of company priorities and interests, was able to patent on occasion. In 1896, for example, he received a patent (No. 334,078) for an "Apparatus for Cooling and Disinfecting" air coming in through a window—not a device using electricity. The same was true of his patent (No. 557,076) for a "Locking Rack for Hats, Coats, Umbrellas, &c" issued in that same year. Clearly, though, his rate of invention had slowed down.

When the General Electric/Westinghouse Board of Patent Control was ended in 1911, Latimer faced a major career decision. His papers in the Queens Borough Public Library contain a drawing he made titled "My Situation as It Looked to Me in 1912." It depicts Latimer on a knife's-edge path balancing himself with a long drafting pen. On one side is "the wide wide world," and on the other is the "General Electric Co. feathers." Ahead along the narrow path is "E. W. Hammer, consulting engineer." Across the bottom is written "which way will he fall?" According to historian Rayvon Fouche, all three choices—staying comfortable with Edison, striking out on his own, or taking a job with the patenting firm Hammer and Schwartz—were

"My Situation as I Saw It," drawing by Lewis Latimer.
Source: The Queens Borough Public Library, Archives, Lewis H. Latimer Papers

appealing to him. Latimer decided on the last option and remained with the firm until his illness in 1924, four years before his death.

At the start of 1918, a call went out to all those who had been with Edison before 1885, proposing the organization of a body to be called the Edison Pioneers. It was an elite group of individuals who had been at the right place at the right time, mastering and advancing a new and powerful industry. Most, like Latimer, would not have been university trained in electrical engineering, but working individually, and in concert had become what was called at the time "electricians." His invitation to join the Edison Pioneers was a signal honor for a black man in a white racist society.

13   George Eastman and the Coming of Industrial Research in America

Reese V. Jenkins

Corporate leadership in modern industrial society is closely linked to research in technology and science. Despite heavy investments in technical research by universities and government agencies, industrial research is of paramount importance to the continued success of many of the largest and most progressive corporations. In general, the development of industrial research facilities and enormous corporate enterprises have gone hand in hand in America. The pioneer industrial research laboratories, including those of Edison, General Electric, Bell, DuPont, and Eastman Kodak, emerged during the late nineteenth and early twentieth centuries in new, large-scale, technically oriented industries related to the rapidly changing electric and chemical sciences. In the chemical industry, the Eastman Kodak Industrial Research Laboratory, although not the first industrial laboratory in the United States, was an early and quite-influential research center. Late in his career, industrialist George Eastman formally established this research institution in Rochester, New York. Through the writings of its British-born founding director, C. E. Kenneth Mees, this research facility served as an industrial model for nearly half a century after its establishment in 1912. An examination of the background of this pioneer research institution reveals the key personal roles of Eastman and Mees, but equally important, illustrates the complex interaction of those social, technical, and cultural factors that not only encouraged the institutionalization of industrial research but also, more generally, helped to shape modern industrial society.

The story of sustained research activity in the photographic industry in the United States begins, however, not in 1912 but instead more than thirty years earlier. The late 1870s and early 1880s stand as a watershed in the basic character of the photographic industry and, ultimately, the nature of industrial research. Before the late 1870s, photography had passed through two technological phases: that of the direct positive daguerreotype from 1839 to about 1855, and that of the negative-positive, wet collodion processes from the mid-1850s to the late 1870s. Because the photosensitivity of the photographic materials was highly perishable, the photographer had to prepare his photosensitive materials at the time and site of exposure. Consequently, the preparation of photosensitive materials was highly decentralized, performed by each individual photographer in his village gallery, on his itinerant wagon, or in his dark tent out in the field.

The character of production influenced the nature of research and development, especially in the area of photographic materials. Immediately following the announcement of the details of the first commercial photographic process in August 1839, the European scientific community showered its interest on photography. Actively conducting research activities were such prominent figures as the Frenchmen François Arago, Hippolyte Fizeau, and Léon Foucault; the Englishmen Sir John Herschel and W. H. F. Talbot; and the American John Draper. In the late 1830s and early 1840s, photography beckoned as a "frontier problem," promising potential insights into the fundamental nature of matter as well as the interrelationship of such "forces" as heat, light, electricity, and clinic effect. During the 1840s, research reports in photography richly sprinkled the scientific literature.

By midcentury, however, the interest of the scientific community waned. Soon photographic societies formed and began publication of their own specialized journals. From the 1850s to the 1870s, research and development was much more sporadic than before. A large part of the industry consisted of individual photographers, who could devote to research only a few fleeting moments stolen from the busy scramble of their highly competitive picture-taking activities. Even in larger firms such as E. and H. T. Anthony and Company of New York, only occasionally did personnel pursue an improvement or development. Apparently viewing scientific or technological creativity as a *spontaneous* event that happened quite unpredictably, like being struck by a bolt of lightning, the two Anthony Company owners took as a new partner a Frenchman whose prime responsibility was to search Europe for novelties. In other words, he was to scour the Old World for the creative products of the sporadic lightning bolts as they struck the inspired geniuses.

Given this romantic conception of technical creativity and the nature of the industry in America, the Anthonys' approach made commercial sense. Most production was limited to small quantities, and it was highly decentralized. Therefore, substantial research investments in the creation of new products would be difficult to recover. Moreover, most American photographers held antipatent attitudes—attitudes derived in part from the highly controversial patent disputes of Louis Daguerre and Talbot in England. Finally, in a decentralized industry consisting of small firms, strict enforcement of patents was not commercially worthwhile.

In the late 1870s, the advent of the much less perishable dry gelatin emulsion created a veritable revolution in production. With it, the character of research changed. Developed in England, factory-made gelatin dry plates were introduced in the United States in the late 1870s. Although the production facilities were initially quite small, they soon grew into large factories with specialized personnel. The most important of these specialized areas was the emulsion making, where improvements protected as trade secrets soon became critical to the commercial success of a company. Because of the specialization, the emulsion makers could devote substantial time to improving their art, and accordingly, some vigorously sought improvements and innovations in photographic materials and their manufacture. With the shift in production of the highly difficult-to-produce photosensitized materials from the photographer to a specialized manufacturer, interest in photography grew among serious, enthusiastic amateurs. Consequently, the market for materials grew, and the financial rewards of innovations increased. Improvements protected by patents and trade secrets played an increasingly critical role.

In such an atmosphere, another major revolution occurred in photography. A diminutive, socially shy yet confident businessman, Eastman had entered the industry about 1880 as a dry plate manufacturer with the goal of obtaining a large share of the American market by employing patented machinery. By 1883, he observed that his machinery patent did not effectively limit competition, and as a consequence, price competition was sharply cutting his profits. He therefore hired a local Rochester camera maker, William H. Walker, to join with him in replacing the dry plate by creating a new system: roll film photography.

During the first half of 1884, they created a roll film holder for replacement in the plate holder on the back of the camera, developed a paper roll

George Eastman made this self-portrait in 1884, noting, "Made on paper with a soluble substratum developed after transferring."
Source: Eastman Kodak Company

film, and designed special machinery for the continuous production of the photosensitized paper film. The concept of the roll holder was not new. Leon Warnerke had introduced such a scheme in England in the 1870s, but he designed his system only for handicraft style of production, sensitized his film only with the slow-speed dry collodion then available, and did not patent his system. The Warnerke scheme decidedly failed. Nearly a decade later in a review of the Walker-Eastman roll-holder mechanism, Warnerke himself raved about the outstanding characteristics of this substantially improved American version. Yet the roll film proved less satisfactory. The professional photographers, who constituted the overwhelming part of the market, complained of the poor images attained from printing through paper. When Eastman introduced a film that required stripping the gelatin emulsion from the paper backing for printing, the professionals continued to ignore the film system because of the tedious manipulations required.

By 1887, the Eastman company had invested substantially in research and development of this roll film system of photography. Each element of the system—roll holder, film, and film production machinery—had been patented. As by-products of the system, the company began production of photosensitized paper, and in order to create a further market for the paper, inaugurated developing and printing service. Both of the by-products were a success, but Eastman came to recognize that the roll film system itself was a commercial failure. Then in 1887 came perhaps the most crucial moments in the history of photography. Eastman, realizing the failure of his system with the professional market, reconceptualized the market, later remarking, "When we started out with our scheme of film photography, we expected that everybody that used glass plates would take up films, but we found that the number that did this was relatively small and that in order to make a large business we would have to reach the general public and create a new class of patrons."

Therefore, Eastman sought to design a simple-to-operate camera for the amateur, drawing on the company's developing and printing service to provide all the services, including the tedious stripping of the film, for the amateur. He designed the camera—the Kodak camera. He sold it loaded with a hundred-exposure roll of film. After the exposures were made, the amateur photographer returned the loaded camera to the factory for the developing and stripping of the film, producing of prints, and reloading of the camera with fresh film. By reconceiving the market and then redesigning the Kodak system accordingly, Eastman essentially created the mass amateur market for photography, revolutionizing photography and with it the photographic industry. By the mid-1890s, mass amateur photography had caught on. Many times the company's rapidly expanding facilities were unable to meet the demands of the Kodak camera snapshooters. Eastman had democratized image making for the world. At the same time, the increased scale of his production permitted greater specialization of function, and encouraged commitments to specialized research and development activities.

With the new mass scale of production and marketing, including substantial overseas operations, a new phase of business strategies and conceptions of technological innovation emerged. Where during the 1850s through the 1870s the leaders of the American photographic industry had sought new developments wherever they occurred spontaneously, during the 1880s a few researchers, with faith that some new solution or alternative could be found, had sought specific developments and, by patenting every element of their new developments, legal monopolies. The experience of multinational enterprises operating within the legal frameworks at home and abroad demonstrated that such strategies did not fully protect the products of their research and development investment. At home, patent litigation was complex and costly, and sometimes required years to come to a decision. By that time the patented feature might be out of date.

Abroad, the patenting of products in a large number of countries was complex, and full protection could not be ensured. Likewise, litigation required extended periods of time during which full commercial exploitation might be curtailed. Hence, at Eastman Kodak there emerged during the mid-1890s the strategy of continuous technological innovation. Eastman succinctly summarized the motivations behind the establishment of an experimental laboratory:

I have come to think that the maintenance of a lead in the apparatus trade will depend greatly upon a rapid succession of changes and improvements, and with that aim in view, I propose to organize the Experimental Department in the Camera Works and raise it to a high degree of efficiency. If we can get out improved goods every year nobody will be able to get out original goods the same as we do.

The No. 1 Kodak camera (1888), disassembled, was priced at $25 and loaded with enough film for a hundred exposures. After exposure, both camera and film were sent to Rochester, the film was developed and printed, and new film was inserted—all for $10.
Source: Eastman Kodak Company

An advertisement featuring an Eastman slogan—"You Push the Button, We Do the Rest"—popularized the No. 1 camera.
Source: Eastman Kodak Company

Here Eastman clearly states a strategy of annual model change—a strategy dependent on a steady stream of improvements from an experimental laboratory. On a less formal basis, a few selected persons at Eastman Kodak pursued similar research and development in the photosensitized products departments as well.

Prior to the formal founding in 1912 of the industrial research laboratory at Eastman Kodak, the research and development facilities could boast substantial achievement. This included thin, rolled celluloid film, cinematographic film, continuous methods of film production such as coating and drying, acetate safety film, an increased degree and range of spectral photosensitivity of emulsions, and considerable color experimentation, plus a host of major and minor improvements in amateur cameras. Hence, industrial research and development was already an ongoing as well as increasingly important activity in the photographic industry. Yet in 1912, Eastman took a bold new step with the establishment of the formalized industrial research laboratory. This new institution reflected, in large part, the attitudes and ideas of two men: Eastman himself and Mees.

The immediate stimulus for the establishment of the laboratory came while Eastman was touring Europe in late 1911 and early January 1912. He traveled in France, Italy, and Germany, signing negotiated film contracts with major film-producing companies. In Germany, he accepted the invitation of the executives of the Bayer Chemical Company to visit their plants in Elberfeldt. After touring the plants, he was feted at a luncheon. During the repast, he sat next to one of the executives who extolled the company's wonderful facilities, including its research laboratory. The executive boasted of Bayer's equipment and the number of its research scientists. Then he turned to Eastman and asked, "And how many people do *you* have in your research work?" Eastman, apparently embarrassed, had to admit that his company had no formalized research laboratory. He returned shortly to England, determined to rectify this situation. At once he embarked on building such a laboratory in Rochester.

Determined to obtain the most promising photographic scientist as the director and architect of the new laboratory, Eastman received from his major European technical-legal adviser the suggestion of young Mees. Eastman approached the tall, gangly Mees, who had just a few years before graduated with a DSc from the laboratory of William Ramsay at the University of London. Mees was a partner in a small dry plate works in Croyden,

Kodak's first folding camera, the No. 4, was introduced in 1890.
Source: Eastman Kodak Company

Moving pictures became a reality when Thomas Edison (right) used Eastman's flexible film in his camera. This photograph marking their joint achievements was taken in 1928.
Source: Eastman Kodak Company

just south of London. He was one of the world's leading authorities on color photography and photometrics. Eastman invited Mees to Rochester to establish the laboratory. Mees, concerned about the plight of his elder partner, expressed his reluctance to abandon his partner and the fledgling firm. Determined to have Mees as director, Eastman bought the dry plate works and employed Mees's partner in the company's Harrow works.

Mees embraced the Rochester opportunity enthusiastically and within the year had the new research facility in operation. Eastman, now entering on a period of substantial philanthropic activity, charged Mees and the laboratory with the responsibility for the "future of photography." More significantly, he committed himself to the support of the laboratory for ten years without expecting any immediate commercial return. Eastman could confidently extend this generous commitment to basic research, knowing full well that Mees was as keenly attuned to commercial results as to basic research.

Mees brought to the laboratory a view of science as an enterprise devoted to accumulating a large body of experimentally determined facts—an ongoing activity that gradually but inevitably progresses ever closer to the true understanding of the nature of the universe. Having abandoned the Fabian socialism of his student days and absorbed Ramsay's enthusiasm for the application of science to industry, Mees embraced a Baconian perspective that saw scientific knowledge as contributing to society. As Mees then viewed it, instead of redistributing the "pie" as the socialists advocate, the enlarging of it through industrial research was the ultimate solution to the plight of the poor. Accordingly, he saw the prototype of the industrial research laboratory not just in Willis R. Whitney's General Electric laboratory or William Rintoul's laboratory at Nobel but also in Francis Bacon's "House of Solomon," which he described as follows:

A great number of observed facts would be collected, and from them the fundamental processes of nature could be understood. In this way, he believed, it was possible to attain to "the knowledge of Causes and secret motions of things, and the enlarging bounds of Human Empire, to the effecting of all things possible." This was a great vision, a new vision on earth, and a vision that has in some measure been realized. The method that Bacon suggested for carrying out this idea was the organization of a research institute, which he entitled the "House of Solomon" and described in his "New Atlantis." This institute was to be manned by a great company of Fellows to whom Bacon ... allotted specific functions. ... A noble dream, much before its time and greatly overorganized, it led to the idea of cooperation in the pursuit of knowledge.

Eschewing the dependence of research laboratories on scientists of exceptional caliber, Mees held

that much scientific research depends upon the accumulation of facts and measurements, an accumulation requiring many years of patient labor by numbers of investigators, but not demanding any special originality on the part of the individual worker. ... He can make valuable contributions to scientific research even though he be entirely untouched by anything that might be considered as the fire of genius.

Furthermore, Mees sounded the death knell of the earlier heroic view of invention and discovery:

It is, indeed, a matter of doubt how many of the men commonly considered to be of great genius by virtue of some important discovery they have made really possessed any distinguishing ability compared with their fellows who did not have the fortune to make a similarly important discovery.

Mees organized his laboratory accordingly, and established the weekly conference system and various bibliographic tools in order to enhance the cooperation among his staff. The multistoried building housed a large library, several laboratories, photographic galleries, and a small-scale production facility. In 1913, he had a staff of twenty and operated on a little over $50,000; by 1920, the staff had increased to eighty-eight and expenditures to nearly $350,000.

During the first decade and a half, the laboratory contributed new understandings in photochemistry and photometrics. Moreover, it introduced the small-scale production of fine organic chemicals when the United States was deprived of its usual German sources during World War I, a home movie system with safety film, and the first color movies. In addition, it sharply increased the range and degree of sensitivity of the company's emulsions through the discovery of the catalytic effects of sulfur compounds in silver halide emulsions. Truly Eastman found the laboratory not only fulfilling its promise of being responsible for the future of photography but also for the "fortunes of Eastman Kodak."

During his industrial career, Eastman did more than create mass amateur photography and thereby democratize image making for the world. He embodied those ideas and values that—along with many others in American and European industry—encouraged him to turn from the romantic view of the spontaneous and sporadic origins of new technological ideas to gradually adopt the view that prompted the continuous pursuit of technological innovation. The research laboratory, based on cumulative, antiheroic views of the origins of new science and technology, stood saliently as the institutionalization of that continuous pursuit.

At the same time, the emergence of continuous innovation was a business response to the legal complications at home and abroad, further fostered by the growing availability of academically trained chemists, physicists, and engineers. Continuous innovation was thus a culturally shaped response to a certain set of broad social conditions. But equally important, it reflected certain social values inherent in Western culture—specifically the high value given to new and improved possessions. In American culture, for example, despite its egalitarianism, society had set aside its antimonopoly attitudes in the development of new technology. The Constitution provided for a patent system, and by law, the inventor was granted a legal monopoly for a period of seventeen years. Moreover, late in the nineteenth century, Americans lionized their patent-seeking inventors such as Thomas Edison. At the same time, they concretized their antimonopoly attitudes with the passage of the Sherman Antitrust Act of 1890.

As such, the forces behind the development of industrial research are multifold and complex. To understand the founding and evolution of this research activity, which is of such importance today, it is necessary to consider the personal motivations and the ideas of science and engineering, and at the same time study the economic, social, and institutional structures that in turn reflect the values of this society. One of the steadfast values of the past two hundred years in America is that technical change is good. The emergence of industrial research laboratories was one manifestation of that value as it filtered through a complex set of cultural and social structures.

14 Ellen Swallow Richards: Technology and Women

Ruth Schwartz Cowan

Ellen Swallow Richards was an extraordinary woman whose life had a profound impact on the history of American technology and the ways in which Americans lead their daily lives—yet she did not exert her influence in the usual way. She did not invent a new device or find a way to manufacture a device that someone else had invented. Rather, she changed American technology by the force of her ideas and the educational movement that was founded on the basis of them: the home economics movement.

Richards was born in 1842 in a small New England village. Her parents were "middling people," schoolteachers, occasional farmers, and owners of the village store. She was their only child, and they educated her at home. In her teens, she became a schoolteacher herself but did not find the work fulfilling; she had a deep desire for wider acquaintance with the world and a more extensive education. Consequently in 1868, at the age of twenty-five, she entered Vassar College in Poughkeepsie, New York—one of the few institutions for higher education for women in the United States at that time. Most of the students at Vassar were younger than she was and considerably wealthier; she supported herself by tutoring them.

On graduating in 1870, she resolved that she wished to do something scientific and practical with her life, and applied to become a student at the Massachusetts Institute of Technology. MIT had been founded just nine years earlier and had never before admitted a female student. The authorities made a special arrangement that must have been ironically pleasing to Richards: they allowed her to attend classes but refused to let her pay tuition so that she would not be a student officially. Official or not, she was awarded a BS degree in 1873—the first woman to be accorded that distinction at MIT. She remained at MIT for additional study for two more years, serving as assistant to the professor of chemistry, William R. Nichols, who was then engaged in analyses of public water supplies for the Board of Health of the city of Boston. During her postgraduate years, she met and married Robert Hallowell Richards, a professor of mining engineering at MIT.

In the ensuing years, her combined interest in science and the problems of women did not flag. Through her marriage, she came in contact with some of the wealthier members of Boston society and managed to convince some of her new friends to help her in various endeavors. One of these efforts was the Women's Laboratory on the grounds of, but not officially connected with, MIT. Here Richards volunteered to teach the rudiments of science to young women who wished to become schoolteachers. Another favorite project was the Society to Encourage Studies at Home, a series of correspondence courses intended to help homebound women acquire an education. Here Richards was the author of the science course. Yet another endeavor was the New England Kitchen, where the diets of working-class families were studied, and public demonstrations were given to show the women of those families how to prepare less costly and more nutritious meals. Richards functioned in this case as chief nutritionist and occasional lecturer. On her own, she was also working as a consultant in industrial chemistry; utilizing her training with Nichols, she analyzed, among other things, the arsenic content of wallpapers and various adulterants in foods.

In 1882, after considerable prodding from Richards and her friends, MIT decided to admit women as regular students (thus closing the Women's

The Women's Laboratory at the Massachusetts Institute of Technology was headed by Ellen Swallow Richards as a separate facility until women were admitted as regular students at MIT in 1882.
Source: Massachusetts Institute of Technology Historical Collections

Laboratory), and two years later, she was appointed as its first female instructor—in sanitary chemistry—and assistant in the laboratory for the chemical study of sanitation (under the direction of her earlier sponsor, Nichols). From that time until her death in 1911, she regularly taught courses in the analysis of air, water, and sewage, and routinely participated in the research projects of the laboratory. She was thus not only one of the founders of the discipline of sanitary engineering in the United States but also among the first women in the United States to be formally employed as a scientist. Those accomplishments mark her as one of the most unusual women of her time, but the fact that she did all those things while maintaining a stable marriage makes her even more unusual. Among the few women who achieved professional status in the nineteenth century, there were even fewer who were married. Richards was fortunate in being married to a man who encouraged and assisted her in her scientific pursuits. She was also fortunate (at least in terms of her career) in having been childless, as the problems of rearing children while maintaining a career were well nigh insurmountable for most of the women who attempted it in the nineteenth century.

Despite her contributions to the development of sanitation engineering, Richards's influence on American technology really lies in another direction. Roughly at the same time that she began her scientific career, educational reformers throughout the United States were beginning to worry about the need to educate young women in the craft of housekeeping. In earlier times, this need had scarcely existed as most girls learned housekeeping either from their mothers or employers, but by the middle years of the nineteenth century, this traditional pattern of education had started to collapse. America was industrializing. It was opening its eastern doors to immigrants from Europe. Girls who would have worked as house servants a generation earlier were now working in factories. Other girls who would have learned housewifery from their mothers now had mothers who did no housework at all but instead relied on immigrant servants and new supplies of industrial goods. When these girls became housewives in their own right, they often did not have access to the same luxuries, but the schools in which they had been educated eschewed practical training as "undignified." For other young women, the conditions of their housewifery were so unlike those of their mothers' as to make their mothers' teachings useless. Consider, for a moment, the situation of a young woman who was reared in a city but found herself setting up housekeeping on the frontier, or that of a young woman who was reared in a peasant household in Europe yet found herself setting up housekeeping in a tenement in New York. All these young women needed training in housewifery, and the educational reformers began to campaign for the inclusion of domestic science (or domestic economy) in school curricula.

The reformers must have been responding to a felt need because their efforts met with success rather quickly. In the West, domestic science courses were frequently offered at the agricultural and mechanical colleges that were created after the passage of the Morrill Act. Iowa State Agricultural College seems to have been the first, with its courses in housekeeping in 1872 and cooking in 1877. In the East, the pattern was somewhat different because the women's colleges then in existence were not hospitable to practical subjects. In cities such as Boston, New York, and Philadelphia, instruction was offered in "cooking schools" especially created for the purpose. Many of these schools were intended for two different types of students: one set of courses was offered for young, fairly well-educated women who needed instruction in the fine points of running a comfortable home, and another set of courses was directed to the poor, uneducated girls who wanted to get good jobs as servants or were ambitious enough to want to learn the American way of housekeeping. The early proponents of domestic science also discovered that they could easily spread their ideas through the written word. Such works as Catherine Beecher's *The American Woman's Home* (1869) and Fannie Merritt Farmer's *The Boston Cooking School Cook Book* (1896) went through many editions in a few years, spreading the gospel of household engineering far abroad in the land.

Early in her career, Richards became a convert to domestic science; its principles, such as they were, pervaded her private life as well as her professional activities. She tried to organize her own home on efficient and healthful lines—eliminating heavy carpets (which collected too much dust), installing many new devices (a gas stove, vacuum cleaner, hot water heaters, and ventilators) as soon as they came on the market, cooking simple meals while trying to ensure that they were nutritious in every respect. Despite the unusual pattern of her own life, Richards's views about household and family were conservative in many respects. She believed that the home was the center of civilized society, and if properly trained, educated women would be the best homemakers of all; they could understand (as untrained women could not) the ways to make their homes comfortable, efficient, healthful and "democratic" (that is, servantless), while avoiding drudgery and leaving themselves sufficient time for other pursuits:

> The educated woman longs for a career, for an opportunity to influence the world. Just now the greatest field offered to her is the elevation of the home into its place in American life. ... Who is to have the knowledge and wisdom and time to carry out the ideals and keep the family up to these standards? Who indeed, but the woman, the mistress of the home, the one who chooses the household as her profession ... because she believes in the home as the means of educating and perfecting the ideal human being, the flower of the race for which we are all existing; because she believes that it is worth while to give her energy and skill to the service of her country and age.

During the 1880s and 1890s, the goal of training the educated housewife in the precepts of domestic science became the focus of many of Richards's professional activities outside MIT. She campaigned to have domestic science included in the curriculum of the Boston high schools, helped to organize the first department of domestic science in an institution of higher learning in the East (Simmons College in Boston), organized the Rumford Kitchen demonstration at the World's Columbian Exposition in Chicago in 1893, and became a leader of the National Household Economics Association that was founded after the exposition. This association attempted to organize women on a volunteer basis to carry out the kinds of programs that Richards had pioneered in Boston.

The fact that the women who worked under the aegis of the National Household Economics Association were volunteers is particularly significant because it points up Richards's other contribution to the development of the home economics movement. She believed that domestic science should be promoted as a suitable career choice for women who were well educated (particularly in the sciences), but who would not, for one reason or another, become homemakers and mothers. Indeed, she thought that teaching or practicing domestic science would be an ideal career choice for such women because it would not violate the norms for female behavior or threaten the professions that women were barred from entering. Richards worked to create courses in domestic science because she wanted some women to be able to learn the craft and other women to be able to teach it; she wanted to create a new profession for women.

To this end, she invited a dozen women who had pioneered careers in domestic science to meet with her at a summer resort in Lake Placid, New

Ellen Swallow Richards
Source: Massachusetts Institute of Technology Historical Collections

York, in September 1897, to discuss what might be done to encourage the professionalization of their field. This meeting was so successful that it was repeated for ten additional summers until, in August 1908, the group, by now considerably enlarged, announced that it was forming itself into the American Home Economics Association. This association continues to this day to serve as the professional nexus for home economists; in its time, it formalized the new profession and gave it a name. (The earlier volunteer group, the National Household Economics Association, had merged with the Federation of Women's Clubs several years earlier.)

Within a few years, the membership of the association had grown considerably. By 1930, it was possible for young women (and some men) to take undergraduate and advanced degrees in home economics in most of the major universities in the United States, and it was also possible for them to find jobs in the field. During the early decades of the twentieth century, home economists came to be employed by school districts, government agencies, social welfare institutions, and a host of business organizations— to teach, plan diets, promote goods, provide homemaking services, and do research.

In the minds of Richards and her colleagues, the goal of the home economist was to bring the individual household out of the Middle Ages and into the twentieth century by integrating it with the technological world of which it was a part. Women, they thought, could not possibly become valued members of an industrialized society if they continued to do their daily work in an unscientific, unmechanical, inefficient, and irrational fashion.

The work of homemaking in this scientific age must be worked out on engineering principles and with the cooperation of trained men and trained women. The mechanical setting of life becomes an important factor, and this new impulse is showing itself so clearly today for the modified construction and operation of the family home is the final crown or seal of the conquest of the last stronghold of conservatism, the home-keeper. Tomorrow, if not today, the woman who is to be really mistress of her house must be an engineer, so far as to be able to understand the use of machines.

This goal, which Richards enunciated in 1910, was achieved remarkably quickly, in part because American manufacturers began to supply American households with the goods and services that were needed, but also in part because the profession of home economics formed exceedingly quickly and helped to guarantee acceptance of the new products throughout the land. As teachers, the home economists trained their pupils to become consumers; as dieticians, they became major consumers of the new goods themselves; as employees of businesses, they advised on what would be acceptable to homemakers and helped to promote the goods they had advised about; as advisers and helpers of the poor, they strove to assist these people to achieve the status and behavior of middle-class consumers; and as scholars and researchers, they concerned themselves with ways to root out and destroy sources of inefficiency as well as economic irrationality in households. Through all this, few of them seemed to notice that their careers were ironic in much the same way that Richards's had been; part-time housekeepers themselves, they made their livings by trying to convince other women that full-time housework was (or could be) the most rewarding career of all.

This aside, it is nonetheless true that Richards's ideas, as her disciplines propounded them, had a mighty impact on American homes in the twentieth century. Before the turn of the century, the vast bulk of the work in those homes was done by human muscle power, much of it female. In most homes, rich or poor, water was hauled, laundry was scrubbed, furniture was dusted, fires were maintained, food was preserved, gardens were tended, and meals were prepared almost entirely by hand. Within decades, much of that work had either been taken over by machines, or eliminated or profoundly reorganized. By 1930, 85 percent of American homes had been electrified; by 1940, half the population enjoyed the benefits of some kind of washing machine; almost every home had indoor plumbing along with hot and cold running water by 1950; and by 1960, there was almost no one left who cooked on a woodstove or coal-burning one. Housework is still time consuming, and for some women it is still onerous, but it is not onerous in the ways that it used to be. The increasing freedom that American women enjoy from household drudgery is part of the legacy of the home economics movement as is their increasing willingness to accept and pay for machinery in their homes. If Americans as a whole have a love affair with machines, this is at least in part due to the fact that most of them live in intimate contact with machinery from the time that they are small children—and this is also part of the legacy of the home economics movement. The ideas that fired that movement, and indeed the very existence of the movement itself, were part of Richards's legacy.

15 Gifford Pinchot and the American Conservation Movement

Samuel P. Hays

During the three decades between 1890 and 1920, a new spirit of concern arose for the future of natural resources in the United States—a spirit that has been called the conservation movement. Contrary to the spirit of the more recent environmental movement, it did not involve a prolonged belief that the nation was running out of resources or there were fundamental "limits" to the earth. Rather, it constituted both a reaction against the hitherto-wasteful use of resources and the assertion that with new scientific knowledge and new techniques, resource waste could be ended and new heights of material production could even be reached. This phenomenon of the Progressive Era called forth significant expressions of the scientific management movement; its advocates believed that through systematic organization and the centralized direction of resource management, the wasteful practices of the past could be turned into the "efficiency" that would lead to a bountiful future.

Gifford Pinchot, the first chief forester of the United States, was one of the architects of this movement, and forest management came to be thought of as one of the central features of the conservation movement. That movement was, to be sure, far more extensive than forestry alone, for it involved the efficient development and management of the nation's waters as well, and a host of collateral resource problems. But Pinchot, dramatic and flamboyant in personality, close to President Theodore Roosevelt and active in partisan politics, often played a role on center stage. Historians have continued to associate the early twentieth-century conservation movement with him only second to the personality of the president himself. As chief forester, Pinchot dramatized the development of the U.S. Forest Service as central to conservation objectives. It is fitting, therefore, that the subject of the role of science and technology in the early twentieth-century resource conservation movement should focus particularly on Pinchot.

Pinchot was the first professionally trained forester in the United States. After graduating from Yale University in 1889, he studied forestry in Europe. He came onto the American forestry scene in the 1890s, first as the manager of Biltmore, the estate of the Vanderbilt family in North Carolina, where he experimented with scientific forestry practices, and then as adviser to federal officials. Scientific forestry programs arose first in the U.S. Department of Agriculture, under which the Bureau of Forestry was established with Pinchot as its head in 1898. The national forest reserves, as they were then called, however, were under the Department of the Interior, which had long supervised the public lands. Through Pinchot's vigorous political action, these reserves were transferred in 1905 to the Department of Agriculture, renamed national forests, and placed under Pinchot's bureau, which soon was renamed the U.S. Forest Service, with Pinchot as chief forester. He served in that position only five years, but in that short period of time he established the outlines of national forest policy for decades to come.

A large measure of Pinchot's influence came from his role in establishing forestry as a profession in the United States. He inspired a considerable number of young people to become professional foresters. Pinchot was instrumental in founding the Yale School of Forestry, where for many years he was a professor. Throughout the nation, many of his protégés took positions in forestry schools attached to the agricultural colleges and universities as well as in state bureaus and divisions of forestry. He played a major role in the development of forestry journals and magazines, conduct of associations such as the American Forestry Association, and development of professional societies such as the Society of American Foresters. All this, it should be noted, did not come without considerable friction, for Pinchot was an irascible individual who did not suffer willingly those with whom he disagreed. His life, in fact, was strewn with strained friendships as well as intense loyalties. He remained the commanding figure in professional forestry for many years while serving as forestry commissioner and later governor of the state of Pennsylvania. Even after his death in 1946, his memory was invoked more than once in debates over forest policy.

Pinchot appeared on the scene at a time when increasing concern was being expressed as to the future of timber resources in the United States. In the northern and eastern parts of the nation, considerable cutting had occurred during the last several decades of the nineteenth century, resulting in devastated wastelands, repeated forest fires, and abandonment of cutover lands. All this had made an intense impression on many Americans. A variety of activities had testified to a concern for future timber supplies, among them the establishment of the American Forestry Association in 1875. The problem was a simple one: Were not future supplies of wood and timber endangered by such wasteful practices? How could one conduct harvest activities with a greater eye to long-run supply? It was in this context that decisions were made to establish forest reserves from the existing national public lands in the West. These would constitute the nation's future timber supply.

To Pinchot, mere forest reserves were not enough. Scientific forest management in the form of sustained-yield practices should be applied. This was the rather-straightforward concept that trees for wood production should be considered as a crop and should be cut each year only to the extent of the annual growth. A stand of timber could be described in terms of its total annual growth calculated in relation to board feet. The annual cut would then be limited to this amount, thus guaranteeing a continuous or sustained yield of wood—hence the term "sustained-yield forestry." There were many other aspects of scientific forestry as well. There was the need to regenerate forests and especially the recently cutover areas. Forestry programs required the production of seedlings in forest nurseries and then planting in the field. Fire protection was required. The felled trees left by loggers in the woods gave rise to spectacular forest fires—in the Great Lake states, frequently wiping out whole villages—and created stark recognition of the need to control this devastation. Fire prevention and suppression programs were critical elements of modern forestry. And there was the need for timber-stand improvement practices, for thinning the growing forest in order to produce a healthy growing stock for future harvest. To these could be added the need for inventories of standing timber on which sound management practices could be based as well as research ranging from investigations into wood utilization, to "forest influence" (for example, on water supply), to insect and pest control. All this and more came to constitute the elements of scientific forestry.

Let us go back a moment and place this movement for scientific forestry in a wider historical perspective. To some extent, the context is one of a change in the attitude of Americans toward their abundant natural resources. For almost three centuries, from the initial settlements on the

Atlantic Coast in the early seventeenth century through the nineteenth, Americans had thought of their nation's resources as inexhaustible. If some resources were exploited, there were more to come. The timberlands of the East were cut down to be replaced by new supplies from the Great Lake states; when these began to run out, new forests were felled in the southern Mississippi Valley; and after these came the Douglas fir forests of the Pacific Northwest. And the same went for farmlands, minerals, water, and energy. Charcoal from wood was followed by coal, and coal by oil, with each found in greater abundance to the west. The productive wheat farming of the Delaware Valley gave way to that of western New York state, then to the eastern Midwest, and finally to the "grain belt" states of the Dakotas, Nebraska, and Kansas. The technologies of the nineteenth century—the new machines in farming, logging, and mining—only served to quicken the pace of resource exploitation. Substituting machines for human labor, they permitted resource extraction that was increasingly more rapid and extensive.

In the late nineteenth and early twentieth centuries, however, these "extensive technologies" began to be replaced by more "intensive technologies," which emphasized more systematic management and "efficiency." Science and technology came to stress themes far beyond the mere substitution of machines and mechanical power for human labor, underscoring those of efficiency, planning, coordination, and the elimination of waste. The application of technology to resources started to stress long-range planning for continuous production, the use of by-products once discarded as uneconomical to utilize, and the application of principles of scientific management to the organization of resource extraction, processing, and marketing. At times these new developments highlighted the need to check the "wasteful exploitation" of past resource use, but more frequently they emphasized the possibilities for the future that could be created by applied science. They occurred, in fact, amid a set of expanding horizons as to the possibilities of human improvement, of "progress" that could occur by means of the application of systematic empirical inquiry and management control to resource problems.

It was this spirit that came to be embodied in the conservation movement, of which scientific forestry was an integral element. But it was not confined to forestry alone. Equally important was the new perspective that arose from the application of large-scale engineering works to the nation's rivers: the harnessing of water, long flowing unused to the ocean, to "useful" purposes for society. The conservation of water by the construction of reservoirs to be used later for irrigation, water supply, navigation, hydroelectric power, and the reduction of flood damage came to be a major theme of the entire conservation movement. Originating in the U.S. Geological Survey, later appearing in the U.S. Bureau of Reclamation, and culminating finally in the 1930s in the Tennessee Valley Authority, multiple-use river development was a major expression of the new scientific management approach to resources. As time went on, a similar approach was applied in the 1930s to the problem of soil erosion. Techniques were developed to prevent the depletion of this major natural resource—the "capital" on which agriculture depended—and establish the scientific management of soil. All these efforts were described by the term "conservation," and those pertaining to water development and forest management were particularly characteristic of early twentieth-century conservation.

This movement was shaped heavily by a new breed of scientists and engineers, who were applying even more intensive amounts of empirical inquiry, engineering works, and management techniques to the development of natural resources. The core of this group consisted of officials in the Roosevelt administration between 1901 and 1909, and included such men as Frederick Newell of the U.S. Bureau of Reclamation, Secretary of the Interior James R. Garfield, Pinchot, and scientist-philosopher W J McGee. McGee especially gave a heavily philosophical rationale to the movement, stressing classic utilitarian themes of the use of science for the "greatest good for the greatest number" of humankind. Such leaders thought of themselves as substituting "scientific" for "political" approaches to resource management, and bringing modern system and order to previously chaotic and wasteful practices. Centralized, so-called efficient management under the guidance of science and technology constituted the spirit of the movement that they led.

The efficient development theme of the conservation movement was an integral part of Pinchot's concept of scientific forestry, and it is important to emphasize this particular twist that he brought to the management of the National Forests. There were at the time competing ideas as to what forestlands should be used for. In fact, the initial forest reserves in the West were established not so much for protecting future timber resources as for protecting watersheds for urban water supply. In this context, one of the

Gifford Pinchot, as chief forester, boards a light plane to inspect the Sequoia National Forest in California.
Source: National Archives

most controversial issues in forest management at the time when Pinchot became chief forester was the impact of sheep and cattle grazing on western watersheds. Advocates of the early reserves established in the 1890s fully intended grazing to be excluded from those lands because of their detrimental effect on water supply through erosion and sedimentation. By 1900, the adverse reaction of cattle and sheep grazers to such pressures to ban grazing endangered the very existence of the forest reserves.

From the beginning of his responsibility for the forest reserves, Pinchot took a stance underscoring the "economic use" of the forests. It was for this reason that he brought about the change in name from forest reserve to national forest. He chose to emphasize the direct economic benefits of forestlands and their direct commodity use. Pinchot brought a leader of the western cattle industry into the national forest administration to develop a program of grazing in the forests. This innovation, not explicitly provided for in law, but developed by administrative action, Pinchot felt was essential in order to lend political support for the national forest program. But it also was more in keeping with his basic orientation of looking on the forests not in terms of protection but rather commodity production.

Pinchot's stress on the economic use of the national forests had much broader implications than simply to increase the role of cattle and sheep grazing. For it involved him in prolonged and persistent controversy over public land issues, and especially the conflicts between commodity and aesthetic uses. His commodity use orientation led to persistent differences with those who wished to reserve lands from lumbering because of their aesthetic value as parks. In the early twentieth century, the movement to establish national and even state parks competed heavily with the national forest movement, and Pinchot took his stand for the latter and against the former. The national park movement, which had originated with the establishment of Yellowstone National Park in 1872, and which by the early twentieth century had resulted in the creation of a half dozen such parks, involved a management philosophy that standing trees should not be cut but instead allowed to mature and decay as a natural system. To Pinchot and the overwhelming majority of professional foresters of the day, this was a "waste" of resources.

As Pinchot was seeking to secure jurisdiction over the forest reserves by transferring them to the Department of Agriculture, he was also seeking a similar transfer of the national parks from the Department of the Interior to his jurisdiction in order to apply commodity management principles to the parks. But leaders of the national park movement, such as J. Horace McFarland of the American Civic Association and John Muir of the Sierra Club, knew well Pinchot's objectives and bitterly resisted his efforts. In this move Pinchot failed. At the same time, park advocates sought to establish a separate administration for the national parks, which they achieved in 1916, when the National Park Service was established in the Department of the Interior. Pinchot vigorously but unsuccessfully opposed this objective.

These were not minor or idle differences of opinion; they constituted major controversies over the proper use and management of forested lands that emerged sharply in the early twentieth century, and have continued to the present day. Pinchot set the tone of the forestry profession. He opposed the establishment of individual national parks in the West, a number of which were carved out of national forestlands; he objected to the protection of the Sequoia trees in California; and he sought without success to open the largest natural park in the East, the Adirondack State Park in New York, to commodity timber cutting. In developing his position, he contrasted the conservation approach of "wise use" of the forests with the preservation approach of those who emphasized forests as aesthetic resources. Down through the years, the Forest Service has fought a rearguard action against aesthetic use, objected to carving national parks out of national forestland, and accepted only with continued reluctance a program of designated "wilderness areas," where commodity timber management is not permitted. It is not too much to say that the association of wise use with commodity use and equally strong reluctance to accept fully aesthetic use—let alone a wise use—on the part of the Forest Service as well as the forestry profession at large was established by Pinchot in the early twentieth century. A more accommodating attitude might well have led to the national forests and national parks being administered by the same federal agency.

Pinchot's drive for scientific forestry also stood apart from another growing "use" of the forests that increasingly came into conflict with commodity production: outdoor recreation—that is, fishing, hunting, hiking, and camping. After the widespread extension of automobile ownership along with the use and development of improved roads in the 1920s, outdoor recreation expanded rapidly in the United States. Much of this took place on public forestlands. And much of it did not fit well with the forestry profession's heavy emphasis on the commodity use of the forests. Outdoor recreationists came into conflict with commodity production in a number of ways. There was the initial competition between game animals and livestock for available forage; Pinchot's decision to legitimize grazing for cattle and sheep on the national forests was a deliberate choice to reduce wildlife objectives in forest management. There were the problems of management for camping and its adverse impact on smoothly run forest protection activities; for instance, careless campers were allegedly responsible for a number of forest fires. And fishermen protested that timber cutting interfered with their sport by reducing water quality and damaging the aesthetic quality of the forest. Most often, contention resulted from simply diverting scarce forest management funds and personnel to outdoor recreation when they could have been used for the scientific wood management programs preferred by foresters.

It was not that professional foresters did not enjoy the out of doors. Pinchot was an inveterate woodsman. But those in the hunting and fishing fraternity often found that their use of the woods did not jibe with the plans of professional foresters. Foresters, in turn, considered outdoor recreationists to be bothersome rather than important and significant users of the forest. It was not until after World War II that outdoor recreation secured a more "legitimate," fully recognized place in U.S. public forest management, and even then, despite the widespread public use of the forests for recreation, it seemed to take a back seat to more predominant wood production uses.

Interestingly, President Roosevelt, a big game hunter and outdoorsman in his own right, had continuous contact with many of the outdoor leaders of the time, and they were also personal friends, such as George Bird Grinnell, editor of *Field and Stream* magazine. Some historians argue that Grinnell was more influential with Roosevelt than Pinchot. That may be. Yet on a number of critical issues, such as the conflict between livestock and wildlife in the national forests, Roosevelt and the big game hunting organization, the Boone and Crockett Club, which he helped to found, were persuaded to keep silent in order to work out a political compromise with the sheep and cattle interests. Outdoor recreation, in fact, became an integral part of forest management and shaped its development far more rapidly on state forestlands than on federal.

One can sense in these remarks some implications of the early scientific forestry movement and conservation for the present day. Most significant is the way in which the term "conservation" has given way to the term "environment"—and not without reason. For while the conservation movement was concerned primarily with the manner of resource development, with a stress on efficiency and centralized direction, the environmental movement has been concerned primarily with "quality of life," and specifically the "quality of the environment" surrounding the home, work, and play of Americans. The change reflects a host of so-called postindustrial values—those associated with people's nonoccupational world as opposed to how they make a living, which are values dealing with home and recreation more than with a job. Such value preferences reflect levels of personal and social income that provide adequately for both necessities and conveniences, and permit special focus on amenities.

The environmental age has several implications for our brief analysis of the role of Pinchot in the early twentieth-century forestry and conservation movements. First is the more general observation that the tendencies toward scientific management at the heart of the conservation movement are actually in deep conflict with tendencies toward the amenity values of environmental quality. The theme of efficiency implicit in the conservation movement logically tended toward more intensive, large-scale development and the comprehensive application of scientific management.

But now the theme of environmental quality persistently comes into conflict with such tendencies. While multiple-purpose river development, for example, was the epitome of the conservation movement, and the Tennessee Valley Authority its most successful result, the environmental movement emphasizes the amenity values of free-flowing streams and opposes dam construction by both the TVA and U.S. Army Corps of Engineers. What was once the triumph of the conservation movement has now become a major target of environmental attack.

Similar tendencies characterize modern scientific forestry. When the major emphasis of forest management was protection, reforestation, and selective cutting, the U.S. Forest Service, despite its controversies with the National Park Service, was held in high esteem. But more recently there has occurred an increased stress on "environmental forestry": the use of the forest for its amenities as an environment for home, work, and recreation. Such a view is in sharp conflict with the extension of scientific forestry in our own day, with its emphasis on increasingly intensive production, genetic selection of seed, fertilization, row planting and cultivation, the use of pesticides and herbicides, and clear-cutting. Controversies between conceptions of the forest as an environment versus a source of commodities are intense; although they are most sharply etched in the disagreements over designating wilderness areas to be left untouched by foresters and other developers, they happen also in the management of the general forest. While intensive forest management represents the tendencies that Pinchot set in motion, those represented by the national parks movement and Muir are closer to the historic roots of the modern environmental forestry movement.

And what of the role of science and technology in all this? Those concerned with commodity production are apt to associate science and technology exclusively with their concerns, and dismiss environmental quality as a rejection of scientific or perhaps any management practices. They complain that by wilderness designations, they are prevented from managing the forest. Yet this confuses means and ends. The dispute is not over the fact of management per se but instead over the goals of management. Are we to stress commodity or environmental quality goals in forest policy? Science and technology are just as essential to the latter as to the former. Their shape and form are different, and they emphasize systematic data collection and application more than machines, and "software" more than "hardware." But even with "wilderness management," an increasingly used term, it is necessary to understand in detail the patterns of resource use and their impact on the ecological system of the forest. Management in this case is geared more to "people management" than commodity management, with an eye to identifying the carrying capacity of the forest and restricting human use to those limits.

Pinchot was a pioneer in the twentieth-century American conservation movement. The current environmental movement is shaped not by tendencies that he helped to set in motion but more by the new values of the post–World War II years and involves impulses that are contradictory to the objectives highlighted by Pinchot. Despite all this, he remains an increasingly legendary figure in the history of American natural resource policy, and especially is associated with the emphasis on harnessing science and technology to large-scale forest commodity management in the United States.

16 Frederick Winslow Taylor and Scientific Management

Gail Cooper

Even as a child Frederick Winslow Taylor was obsessed with order, making up elaborate rules for the informal games his playmates organized. That search for order characterized his working life, but it also touched the concerns of the nation in the decades spanning the turn of the century. Immigration, urbanization, and industrial growth were revolutionizing the traditional American society of small towns and personal relationships. Along with these changes came social strife. Taylor, a mechanical engineer, attempted to find technical solutions to labor conflicts in American industry by increasing the surplus of wealth through greater efficiency. His study of order in the workplace prompted a popular craze throughout America for both personal and national efficiency. Theodore Roosevelt once stated, "The individual is first made efficient. ... From these efficient units is built up an efficiency organization. And when we get efficiency in all our industries and commercial ventures, national efficiency will be a fact."

Taylor's system for greater industrial efficiency, called scientific management, was aimed at increasing national productivity as well as eliminating the conflict between workers and employers. As he explained to a congressional committee in 1911,

The great revolution that takes place in the mental attitude of the two parties under scientific management is that both sides take their eyes off the division of the surplus as the all-important matter, and together turn their attention toward increasing the size of the surplus until this surplus becomes so large ... that there is ample room for a large increase in wages for the workman and an equally large increase in profits for the manufacturer.

While he expressed sympathy for those who were overworked in grimy sweatshops, Taylor believed workers who underworked posed a much greater problem. He accused them of "soldiering," deliberately restricting their speed and output. Taylor argued that if workers increased their production, the resulting surplus would be sufficient to give workers a greater daily wage and management larger profits. This fair day's wage for a fair day's work would benefit both workers and employers. Taking a fairly simple view of human motivations, Taylor believed scientific management gave workers what they chiefly wanted—high wages—and employers what they most desired—a low labor cost. Rather than merely exhorting employees to work harder, management should employ "scientific" methods to establish a standard for a proper day's work.

Taylor was right that some workers deliberately limited their production. Workers called this a "stint," the amount that they had informally determined to be a fair day's work. The stint was an average of the workmen's abilities and protected them from unreasonable speedups by the boss. To exceed the stint for the sake of better pay or to please the boss was considered selfish and simply a way of advancing an individual's self-interest at the expense of his buddies. Workers characterized those who did so as "hoggish" or "unmanly." Whether the stint represented a reasonable amount of work or not was irrelevant to Taylor. Fundamental to his system was the claim that management, not workers, should control the workplace. This conflicted with traditional work patterns in many industries where skilled craftsmen supplied the technical know-how to run the workshop and management concentrated on the commercial side of the business. In these industries, craftsmen enjoyed a large degree of autonomy over the techniques of production, and even hired and paid their own assistants. That monopoly over technical knowledge gave skilled workers considerable power over their own working conditions. In contrast, Taylor contended that management should control the methods of production and there was no limit to management's right to know.

Taylor drew on his own experience as both a workman and manager to give credibility to his system. Born into a wealthy Philadelphia family in 1856, Taylor was expected to become a lawyer like his father and, in 1874, passed the Harvard University entrance examination with honors. He must have felt some ambivalence about his family expectations, though, for the same year he began to complain of failing eyesight, abruptly withdrew from school, and took a position as a journeyman machinist. Although his eyesight was soon "restored," Taylor continued training as a machinist, earning nothing his first year, $1.50 a week in the second and third years,

Frederick Winslow Taylor
Source: Samuel C. Williams Library, Stevens Institute of Technology, Hoboken, New Jersey

and $3 a week in his fourth. However modest his position at work, Taylor never totally rejected the privileges of the class he was born into. During the four years of his apprenticeship, he played cricket at the Young America Cricket Club, sang with a choral group, and participated in amateur theatricals like many young men of the Philadelphia upper class. When he joined the Midvale Steel Works in 1878 as a common laborer, one owner, E. W. Clark, was the father of his tennis partner; the second, William Sellers, was a family friend whom Taylor always addressed as "Uncle William." While at Midvale, Taylor took a home course in engineering and graduated from Stevens Institute of Technology.

It was at Midvale that Taylor was first appointed to a supervisory position as gang boss of the machinists. As a machinist, Taylor had worked the stint like the others; as a supervisor, he was determined to increase production despite worker resistance. "Those men were my personal friends," Taylor said later, "but when we went through the gate of that place we were enemies." The antagonisms between labor and management were intense throughout American industry, fueled in part by the obvious contrasts between the personal fortunes amassed by a few industrialists like Andrew Carnegie and grim conditions of the urban working poor. Middle-class reformers who were part of the Progressive movement brought public attention to poor working conditions in muckraking exposés and fictional works like Upton Sinclair's *The Jungle*. But for some people, sympathy turned to fear as workers unionized and strikes occasionally led to violence. While workers could often count on the sympathy of local law enforcement officials in their communities, large companies could call on state and federal government troops to break strike efforts. The effect was of a nation at war with itself; national concern about "the labor problem" led to the creation of the Department of Labor in 1913.

By the turn of the century, an increasing number of engineers were finding themselves in management positions mediating between the interests of workers and industrialists. Both the growing scale and technical complexity of industrial concerns pushed the formation of a professional managerial class. The railroads pioneered modern management techniques when they adapted the army's system of line-and-staff organization to coordinate their far-flung enterprises. In 1886, Henry R. Towne suggested to the American Society of Mechanical Engineers (ASME) that engineers could make important contributions to the better management of industry. His idea was timely; in the forty years between 1880 and 1920, the engineering profession increased by almost 2,000 percent, from 7,000 to 136,000 members. It is not surprising that Taylor first presented a part of his system of scientific management to the society in an 1895 paper titled "A Piece Rate System Being a Step toward a Partial Solution of the Labor Problem."

Taylor's system included a reorganization of the factory, timed study of individual tasks, and piece rate rather than hourly wage. Work was routed through a central planning office responsible for supplying raw materials and the appropriate tools along with directing each individual task. It was no longer part of a machinist's job, for example, to locate his own materials, or clean and sharpen his own tools. Once the factory was rationalized, efficiency experts used stopwatches to time workmen at each job, stopping the clock to exclude any unnecessary or repetitive actions, and establishing the most efficient time and method possible for each task. Workers were given a written instruction card on how to do each task. A differential piece rate rewarded those who exceeded the standard by paying more for each piece produced than was paid if the old level of production was maintained, and under Taylor's original plan, those workers who failed to produce up to the standard were paid at such a low piece rate that they were encouraged to quit and find other work.

By standardizing working conditions and timing each task in the factory, Taylor believed he had turned management into a science. He argued that there was "one best way" to do every kind of work, and that it was best determined by management. Workers were incapable of directing their own work "either through lack of education or through insufficient mental capacity." Critics charged that by separating the planning of each task from its execution, Taylorism turned workers into virtual human machines.

That attitude toward workers can be seen in Taylor's attempts at improving efficiency at the Bethlehem Steel Company. Taylor left the Midvale Steel Company in 1890 to become a management consultant. After several unsuccessful jobs in which he argued with top company officials over the extent of the engineering expert's power in the factory, a former supervisor at Midvale paved the way for Taylor to come to Bethlehem Steel Company in 1898 to institute a piece-rate system. Taylor took with him several of his experienced associates including Carl Barth; although stormy, the years at Bethlehem were fruitful for Taylor. He combined a reorganization of the steelworks with experimental tests of his idea about labor. Traditionally machinists thought the best method for running the lathe was one that preserved expensive cutting tools for the longest useful life. Taylor's goal was to find a method that removed the most metal in the shortest time. He conducted systematic tests on machine lathes, and discovered that standard steel-cutting tools that contained small amounts of tungsten and chromium could remove 200 to 300 percent more metal when run at high speeds that heated the cutting steel to near the melting point. With Barth's mathematical skill, Taylor assembled slide rules that told machinists exactly how to set their machines for each cut. This discovery of "high-speed steel" earned Taylor considerable attention in engineering circles, and in 1906, he was elected president of ASME. His experience with high-speed steel also supported Taylor's conviction that management, not the workforce, was best able to control even skilled work in the factory.

Other tests on simple tasks reinforced his belief. At Midvale Steel Works, Taylor's compulsive systematization led to years of unsuccessful experiments on ways to measure work effort. "The greatest obstacle to harmonious cooperation between the workmen and the management lay in the ignorance of the management as to what really constitutes a proper day's work for a workman," Taylor wrote. Barth's mathematical analysis of the data convinced Taylor that maximum work could be achieved by regulating the percentage of time a worker spent at work and rest. In one of the simplest jobs at Bethlehem, workers loaded pig iron, rough cylinders of iron, onto a railroad car for ten hours a day at the rate of 12.5 long tons of iron per man per day. Each man picked up a 92-pound pig, walked up an inclined ramp, and dropped the chunk of metal onto the bed of the car. From his theories on rest and work percentages, Taylor calculated that a first-rate pig iron handler was capable of loading 47.5 long tons of pig iron (106,400 pounds) a day, but that seven out of eight men would not be physically capable of such work. Selecting the proper workmen for each task was important to Taylorism. Equally significant was bargaining with individuals rather than groups of workers, for Taylor wanted to substitute individualism for the group ethic represented by the stint.

For his experiment, he singled out a man he called Schmidt, who had ambition and stamina, and often trotted home to then work on a house he was building for himself on a small piece of ground that he had bought with wages of $1.15 a day. Taylor knew no worker would agree with a supervisor's suggestion that with more effort, he could increase his work from 12.5 to 47.5 tons per day, so he illustrated the kind of approach that was necessary with workers. He took Schmidt aside, and after some preliminary questions told him, "What I want to find out is whether you are a high-priced man or one of these cheap fellows here. What I want to find out is whether you want to earn $1.85 a day or whether you are satisfied with $1.15, just the same as all those cheap fellows are getting." When Schmidt agreed he wanted to earn $1.85 a day, Taylor indicated a pile of iron and told him that to earn his higher wage, he would have to load all of it the next day under the direction of an efficiency expert. Taylor informed him roughly,

Well, if you are a high-priced man, you will do exactly as this man [pointing to the time-study man] tells you tomorrow, from morning to night. ... Now a high-priced man does just what he's told to do, and no back talk. Do you understand that? When this man tells you to walk, you walk; when he tells you to sit down, you sit down, and you don't talk back at him. Now you come on to work here tomorrow morning, and I'll know before night whether you are really a high-priced man or not.

Under that kind of supervision, Schmidt loaded 47.5 tons of pig iron, earning 60 percent higher wages than ever before for 380 percent more work. Taylor saw no inequity in this development, focusing instead on the fact that both the workman and company were better off. Nor did he have any regrets about Schmidt's loss of control over even the minutia of his job. Rather, Taylor felt confirmed that he had discovered a new science. In his study he concluded,

Therefore the workman who is best suited to handling pig iron is unable to understand the real science of doing this class of work. He is so stupid that the word "percentage" has no meaning to him, and he must consequently be trained by a man more intelligent than himself into the habit of working in accordance with the laws of this science before he can be successful.

Taylor's hopes to establish the engineer as the practitioner of the new science of management with benign dictatorial powers in the factory inevitably collided with the power of traditional management. At the Bethlehem Steel Company, Taylor's insistence on total control over production came at the expense of both workers and top management, and in 1901 he was fired. Yet the public found his ideas of solving the "labor problem" by "scientific impartiality" to be powerfully attractive. After leaving Bethlehem Steel, Taylor retired from engineering and devoted himself to the promotion of his system of management. Interest in his system spread beyond industry to the general public in 1910 when attorney Louis Brandeis argued at a rate-increase hearing that inefficient management by the railroads ultimately cost the consumers and the adoption of scientific management could save the railroads $1 million a day. Such a sensational claim captured the public's attention. In 1911, Taylor published *The Principles of Scientific Management* to explain scientific management, and his admirers formed the Taylor Society in 1912 to perpetuate his work. Many of Taylor's followers and associates, including Barth, Morris L. Cooke, Henry L. Gantt, and Frank and Lillian Gilbreth, became prominent figures in the efficiency mania that swept the nation.

Taylor's concern with efficiency was given a slightly different twist by the Gilbreths, who worked as a team to introduce their own brand of scientific management to industry. Frank Gilbreth (1868–1924) had begun his working life as a bricklayer, and his interest in speed and efficiency led him to a study of motion. He began with the bricklayer's craft, inventing special scaffolding at waist height that eliminated the need for bricklayers to bend over to pick up each brick and at the same time teaching them to work with both hands. He became a successful contractor in Boston, and in 1904, published *Field System*, a description of his methods. After 1912, he gave up contracting, and together with his wife, devoted his energies to motion study and scientific management.

The Gilbreths's contribution was to break down work into basic motions and eliminate unnecessary movements. Frank Gilbreth named these basic motions "therbligs," which he found unnecessary to point out was Gilbreth spelled backward. His analysis of movement became so detailed that he used motion picture cameras with a finely graduated clock appearing in

Lillian Gilbreth poses for a study of movement.
Source: National Museum of Natural History, Smithsonian Institution

Frank and Lillian Gilbreth set up a time-and-motion study of a clerical worker.
Source: National Museum of Natural History, Smithsonian Institution

the lower corner to record the movement of workers. Unlike Taylor, the Gilbreths's goal was not only to save time but also to eliminate worker fatigue so as to "make it easy for a man to work hard." Frank Gilbreth found Taylor's scientific management to be congenial to his own interests, and his personal admiration of Taylor was spoiled only when Taylor refused to recognize micromotion studies as different from his own time study.

Frank Gilbreth's death in 1924 left his wife, Lillian Moller Gilbreth (1878–1972), as the sole support of their eleven surviving children. As a partner in the Gilbreth consulting firm, Lillian's strength was the use of psychology in management. An Oakland, California, native who was the first female graduation speaker at the University of California at Berkeley, she earned a doctorate in psychology from Brown University in 1915. As a woman in engineering, she encountered some barriers to full participation in her profession. She became an early member of the newly founded American Management Association, but despite her active involvement in industrial engineering, the more traditional American Society of Mechanical Engineers twice barred her from entering the all-male Engineering Club in New York to attend meetings. Her always-busy schedule combined with the Gilbreth enthusiasm for efficiency meant that their large household became a laboratory for domestic efficiency measures that were humorously portrayed in the book *Cheaper by the Dozen*.

After her husband's death, Lillian Gilbreth continued to consult with corporate clients, although she turned more and more to teaching management techniques, and joined the faculty at Purdue University in 1935. She devoted considerable time to domestic efficiency. Her emphasis on an improved physical layout and the elimination of fatigue in the home made substantial contributions to the field of home economics. Most attempts to apply Taylor's scientific management to the household, however, foundered on the substantial differences between the home and factory as a workplace. Separating the planning and execution of each task was a questionable exercise when the housewife represented both labor and management, selecting the proper workman for each task futile when the housewife was solely responsible for a wide variety of chores, and motivating workers with piece rates pointless as long as housework remained an unpaid occupation. Recommendations to apply scientific management to the home urged housewives to save time by more efficient planning and then use that time for individual pursuits, just as workmen were encouraged to excel at their jobs for individual gain. But most women appear to have reinvested any spare time gained into household tasks that would raise the standard of living for their families.

While Taylor's primary interest in efficiency was to eliminate soldiering, those who followed him answered the question, "Efficiency by what measure?" in a different way. Gantt (1861–1919) was an engineering graduate of Stevens Institute of Technology who worked with Taylor at both Midvale and Bethlehem. He was a southerner by birth and tradition whose lifelong hero was Thomas Jonathan "Stonewall" Jackson. The Civil War had created financial disaster for Gantt's family, and he was educated at a military-style boarding school for poor boys just outside Baltimore. He first met Taylor at the Midvale Steel Works. In 1987, Taylor attempted to reorganize the Simmonds Rolling Machine Company but was met with strong resistance by middle management. When the general manager, all foremen and assistant foremen, the salesmen, and the head men in the office resigned on three days' notice, Taylor installed Gantt as superintendent. From Simmonds, both men went to the Bethlehem Steel Works.

Gantt was temperamentally kinder and more flexible than Taylor, and at Bethlehem he modified Taylor's differential piece rate with the task-and-bonus system. Under this system, workers who failed to achieve their task for the day were simply paid by the hour rather than at Taylor's low piece rate, while "efficient" workers were paid a bonus. Gantt's later willingness to compromise with his engineering clients made Taylor wonder if Gantt really understood scientific management, and like his relationships with others, Taylor's friendship with his protégé soured after 1908 when Gantt developed independent ideas about workers' training. As his thinking about industrial management evolved, Gantt was as critical of inefficiency in top management as Taylor was of labor. His most important technical contribution, a graph showing machine production known as the Gantt chart, became part of his attempt to establish standards for management analogous to time studies for workers.

When workers failed to achieve their bonuses, Gantt was always interested in finding out why; his investigations often led him to discover a failure in management. His critique of management shifted from a concern with increased profits to a deepening conviction that factories should be run for the public good instead of merely for private profit. Just as Taylor aimed to benefit workers and employers by maximizing production and profits, Gantt extended that concern to include economical production for the benefit of the consumer. He deplored the tendency of inefficient companies to allow the machinery to stand idle and pass the cost of this expense onto the customer. Industry worked at only a small percentage of its capacity because shortages resulted in higher prices and higher profits for factory owners. To discourage the industrial capacity of the country being run for the private profit of a few, Gantt advocated a new system of bookkeeping in which the cost of idle machinery was not added to the product but instead taken directly out of profits.

In contrast to his disillusionment with the selfish concerns of financiers and businessmen, he envisioned an important role for the engineer not only in industry but also in society. Influenced by economist Thorsten Veblen, at a meeting of ASME in 1916, he and thirty-three others formed a group called the New Machine, dedicated to increasing the purchasing power of a day's pay in New York City by readjusting the social and political approaches to the challenges of industrial society. Gantt, however, did not have the personality for politics. As his biographer observed, "His inherent honesty, lack of tact, fearless straightforwardness, brusque expression, and loneliness, because of lack of friends, are not the human qualities which make for the success of the politician." The New Machine was pushed aside by the concerns of World War I.

While many of Taylor's followers expanded both the techniques and scope of his system, Barth remained the most orthodox of the scientific management experts and was one of Taylor's favorite lieutenants. In 1909, Taylor recommended that Barth bring scientific management to the U.S. Federal Arsenal in Watertown, Massachusetts. The arsenal became a battleground between exponents of scientific management and organized labor. Taylor was anxious to gain implicit approval of scientific management from the federal government, while the union's resistance focused on preventing the introduction of the stopwatch. When Barth's assistant conducted his first time study in the foundry, workers surreptitiously timed the man under study as well; the resulting discrepancy fueled a strike the next day among the foundrymen. The expert was timing an ideal process, not the actual work time, but had to admit that he knew little about the technical requirements of the task and had settled on a time based on what he felt was appropriate.

The "science" of Taylor's system seemed ultimately to have been based on subjective judgment cloaked in precise measurement, and even supporters who believed the system to be effective resisted its claims to scientific precision. When Taylor asked the American Society of Mechanical Engineers to form a special committee to evaluate the scientific basis of Taylorism in 1912, it returned a report against Taylor citing the imprecision of terms like "first-class workman" and the substantial allowances that had to be added to each task time to account for unavoidable delays. In reality, Taylor was trying to measure the unmeasurable, and the intolerance of his "one best way" was a measure of the man as well as the system.

After congressional hearings on the new system, workers succeeded in blocking Taylorism at Watertown Arsenal by lobbying their congressmen to outlaw the use of Taylorism on some government contracts. The passage of congressional legislation against the use of scientific management left Taylor discouraged. He lost heart for his work, and his health declined. He suffered from a lingering cold; on a trip to visit his sister in Philadelphia, his cold turned to pneumonia, and he died on March 21, 1915, at the age of fifty-nine.

Taylorism was only a partial success during its founder's lifetime due to Taylor's difficult personality and the inherent flaws in the system. Scientific management was seldom adopted in its entirety, and the comprehensive installation at the Tabor Manufacturing Company, which Taylor used repeatedly as a living example of his methods, was notable for its rarity as well as effectiveness. In addition, claims of scientific precision crumbled during Taylor's lifetime, and with them the force of its moral imperative. Taylor had maintained that the discipline of the task-and-bonus system improved the moral character of the worker, and led to sobriety, ambition, and thrift. That view of the relationship between production and workers' welfare was inverted by a more persuasive group of reformers who provided workers with subsidized housing, savings plans, and leisure activities in the belief that material well-being was the precursor to increased production.

Yet the highly publicized victory of labor over Taylorism did not stop the spread of scientific management. Taylorism was remarkably influential both at home and abroad, where it seemed to many to be an integral part of America's industrial success. In addition, scientific management experts like Lillian Gilbreth launched the new field of industrial psychology and personnel management. Scientific management was also long lived. The increasing technical complexity of large-scale industry from 1880 to 1920 that encouraged the growth of systems of management remained a permanent feature of industrial America.

17 Henry Ford and the Triumph of the Automobile

James J. Flink

By 1908, the year in which Henry Ford introduced his Model T and William C. Durant founded General Motors, the automobile, in the words of technology historian John B. Rae, "was already potentially what it would become in fact—an incredible item of mass consumption." By 1927, when the Model T was withdrawn from production, the automobile industry was America's leading industry in value of product; there was an automobile registered for every 4.5 persons in the United States, and 55 percent of American families owned cars. A presidential commission surveying contemporary changes in American life concluded about the automobile in 1933, "It is probable that no invention of such far reaching importance was ever diffused with such rapidity or so quickly exerted influences that ramified through the national culture, transforming even habits of thought and language."

More recently, Father R. L. Bruckberger, a French observer of American life, has concluded in his *Image of America* that Ford's innovation of the Model T, moving belt assembly line, and five-dollar, eight-hour day have had more important consequences than Vladimir Lenin's socialist revolution. Currently the automobile industry directly provides one out of every six jobs in the United States; there is an automobile registered for every 2.25 persons, and 83 percent of American families—all except the poor—own cars. The automobile makes possible and has come to symbolize much of the American way of life.

Ironically, "the automobile is European by birth, American by adoption." Until shortly after the turn into the twentieth century, Europeans were about a decade ahead of Americans in the development of the gasoline-powered car. The automobile was pioneered to the stage of commercial feasibility as early as 1887 by Karl Benz, a German manufacturer of stationary gas engines, and the mechanical arrangement of the modern motorcar was innovated in 1891 by the French engineer-entrepreneur Emile Constant Levassor. In 1895, Levassor drove one of his cars over the 1,170-kilometer course of the Paris-Bordeaux-Paris race at the then-incredible speed of just over 24 kilometers per hour, with the longest stop for servicing being only twenty-two minutes. This demonstrated that the replacement of the horse by a motor was more than an idle dream and touched off a flurry of automotive activity in the United States, where Levassor's achievement was given extensive coverage in the newspapers. Although a handful of American inventors had managed to build experimental cars by the mid-1890s, their machines were primitive by European standards and were not being offered in quantity for sale. In contrast, by 1895, several French and German automobile manufacturers were issuing catalogs, and automobiles were already a common sight in the streets of Paris.

Charles E. and J. Frank Duryea, two Springfield, Massachusetts, bicycle mechanics credited with building the first successful American gasoline-powered car in 1893, initiated the commercial manufacture of motor vehicles in the United States in 1896. Other early experimenters, including Ford, soon entered the competition, and some thirty American automobile manufacturers produced an estimated twenty-five hundred motor vehicles in 1899, the first year for which separate figures were compiled for the automobile industry in the *United States Census of Manufactures*. Despite the fact that cars were then sold on a cash-on-delivery basis and left much to be desired in performance, the demand for motorcars in the United States far exceeded the supply. The early automobile manufacturers operated in an unprecedented seller's market for an expensive item. Consequently, by 1910, automobile manufacturing had leaped from 150th to 21st in value of product among American industries, and had become more important to the national economy on all measurable criteria than the wagon and carriage industry. By 1910, 458,500 motor vehicles were already registered in the United States, making America the world's foremost automobile culture.

There are many reasons why, in contrast with Europe, an automobile culture developed so rapidly in the United States. To begin with, the volume production of standardized commodities became well established early in our industrial history. Our abundance of natural resources, combined with a chronic shortage of labor, resulted in low costs for raw materials and the mechanization of industrial processes, which necessitated the standardization of products. In addition, the absence of tariff barriers between the states encouraged sales over a wide geographic area. Most important were our higher per capita income and more equal income distribution relative to those in European countries. It is significant, for example, that Morris Motors, the most important British automobile manufacturer, did not install a moving assembly line until 1934—two decades after it had been innovated at Ford. The income distribution in Great Britain fixed the demand for cars there at too low a level to justify the investment. Because of these differences between Europe and the United States, the European pattern of small-scale, individualized production of motor vehicles stood no chance of becoming characteristic of the American automobile industry.

Americans enthusiastically received the automobile from its introduction, and shortly after the turn of the century, predictions became commonplace that a utopian "horseless age" would dawn with the imminent development of the low-cost, reliable car. Even before the introduction of the Model T in 1908, most Americans anticipated that mass personal automobility would soon become a reality. The bicycle had made Americans conscious of the possibilities of individualized, long-distance highway transportation, and the crest of the bicycle movement in the 1890s coincided with the climax of several decades of agrarian unrest that had singled out as a prime target the abuse of monopoly power by the railroads. Farmers and city consumers alike began to perceive cheaper highway transportation as an alternative. Vast areas of the American continent remained inaccessible by rail or highway, and most Americans still lived on isolated farms or in poorly interconnected rural villages. They wanted better access to not only markets but also the social and cultural amenities that only the metropolis provided, particularly to better medical care and education. In cities, the antiseptic automobile was viewed as an answer to the health problems arising from ever-present manure heaps and the traffic congestion caused by overworked dray horses dropping dead in their tracks. The urban middle class, imbued with the Jeffersonian "agrarian myth," sought escape to suburban residences through automobility. Municipal governments hard pressed to meet the mounting costs of building adequate mass transit rail systems saw an answer in the widespread adoption of motor vehicles. Countless early demonstrations convinced people that the automobile was more economical, more reliable, and safer than the horse-drawn vehicle. And perhaps most important, the motorcar offered our individualistic, migrant population the promise of greatly expanded personal mobility and freedom of choice in residence, business location, and the pursuit of leisure-time activities.

Ford was by far the most successful of the many early automobile manu-facturers who attempted to build a low-cost, reliable car for this developing mass market. Born on a Michigan farm in 1863 of middle-class parents, Ford early on developed a lifelong aversion to the drudgery of farm labor. As a youth, he loved to tinker with machinery, and while employed as an engineer at Detroit's Edison Illuminating Company, he built his first exper-imental automobile in 1896. During two unsuccessful attempts to enter automobile manufacturing, Ford gained a national reputation as a designer and driver of racing cars. Then at the age of forty, with new backers, Ford formed the Ford Motor Company in Detroit, Michigan, on June 16, 1903, to produce cars in the intermediate price range. The Ford Motor Company started in business with paid-in capital of only $28,000, a dozen workmen, and an assembly plant just 250 by 50 feet (76 by 15 meters). So little capital was needed because Ford, like almost all other early automobile manufac-turers, jobbed out the manufacture of components to scores of independent suppliers. Once the design of his car was established, the early automobile manufacturer became merely an assembler of major components and sup-plier of finished cars to his distributors as well as dealers. As a consequence of these easy conditions of entry—low capital requirements combined with the promise of high profits—some 515 separate firms had entered automo-bile manufacturing by the time Ford came out with his Model T in 1908. Competition was intense; firms that made technological blunders or failed to adjust to changes in demand were weeded out with Darwinian ruthless-ness. The vast majority of Ford's early competitors failed. Why did he suc-ceed so brilliantly?

Ford was not the first automobile manufacturer to attempt to produce in volume low-cost, reliable cars. That distinction belongs to Ransom E. Olds, who initiated volume production of his $650 curved-dash Oldsmobile in 1901, while Ford was still building racing cars. But both Olds and Thomas B. Jeffery, who followed Olds into volume production with his $750 to $825 Ramblers in 1902, attempted to mass-produce cars that were soon out-moded. The surrey-influenced design of the curved-dash Oldsmobile had inherent mechanical weaknesses and thus was rapidly abandoned by most manufacturers. The typical gasoline automobile of 1908 bore little resemblance to the horseless carriage of 1900. Numerous improvements meant that the 1908 gasoline automobile was a fairly reliable family car. The problem remained prior to the Model T, however, of making such cars available at prices the average family could afford, by cutting manufacturing costs or reducing unit profits without sacrificing the quality of the product.

At the time the Ford Motor Company was organized, the automobile industry was dominated by the Association of Licensed Automobile Manufacturers (ALAM), a select group of makers of gasoline automobiles who were attempting to monopolize automobile manufacturing in the United States through control of a patent on the gasoline automobile that had been awarded in 1895 to George B. Selden, a New York patent attorney. The ALAM companies (which accounted for about 80 percent of the industry's total output prior to 1908) were in the main committed to maintaining high unit profits through producing high-priced cars for a limited market. The ALAM tried to set production quotas and freeze new entrances into automobile manufacturing. Ford was denied a license to manufacture gasoline automobiles under the Selden patent on the ground that he had not demonstrated his competence, and when Ford persisted in producing cars, the ALAM immediately brought a lawsuit against him for infringement of the Selden patent. The suit was ultimately decided in Ford's favor in 1911, and the ALAM disintegrated. Meanwhile, Ford turned the suit into a clever public relations campaign by contrasting his own status as a pioneer automotive inventor and struggling small businessman with the image of the ALAM as a group of powerful, parasitic monopolists.

Like many other automobile manufacturers who operated without licenses from the ALAM, Ford was increasingly committed to the volume production of light, low-priced cars. His $600 Model N was one of the better-designed and better-built cars available at any price in 1906. The *Cycle and Automobile Trade Journal* called the Model N "the very first instance of a low-cost motorcar driven by a gas engine having cylinders enough to give the shaft a turning impulse in each shaft turn which is well built and offered in large numbers."

Encouraged by the success of the Model N, Ford was determined to build an even better low-priced car. At $825 for the runabout and $850 for the touring car, the four-cylinder, twenty-horsepower Model T was first offered

Henry Ford stands beside the first automobile he built in 1896, while his son Edsel inspects the ten-millionth Ford car, a Model T, which Ford introduced in 1908. *Source*: Ford Motor Company

to dealers on October 1, 1908. Ford's advertising claim about the Model T was essentially correct: "No car under $2,000 offers more, and no car over $2,000 offers more except the trimmings." Committed to large-volume production of the Model T as a single, static model at an ever-decreasing unit price, the Ford Motor Company innovated mass-production techniques at its new Highland Park plant, which permitted prices to be reduced by August 1, 1916, to only $345 for the runabout and $360 for the touring car. Model T production increased form 32,053 units in 1910 to 734,811 units in 1916, giving Ford about half the market for new cars in the United States at the outbreak of World War I.

The Model T was the archetype of a uniquely American mass-produced gasoline automobile. Compared with the typical European touring car, the American-type car was significantly lower priced, was much lighter, had a higher ratio of horsepower to weight, and was powered by a larger-bore, shorter-stroke engine. In addition to the increasingly greater emphasis that American manufacturers gave to producing cars for a mass market, these characteristics of the American car resulted from the lower price of gasoline and absence of European horsepower taxes. The sacrifice of fuel economy for greater engine flexibility made it possible to build cars that could more readily negotiate steep grades and wretched roads, and were easier to drive because they required less frequent shifting of gears. The American car, designed for the average driver rather than the professional chauffeur, was more apt to withstand abuse and was simpler to repair than the typical European car. With a body that sat higher above the roadbed, however, the American car appeared ungainly, and showed less attention to the details of fit, finish, and appointment. That the chassis of a car cleared the invariable hump in the center of our unpaved, rutted roads was then more important to the American consumer than style.

The sixty-acre Highland Park plant that Ford opened on January 1, 1910, to meet the huge demand for the Model T possessed an unparalleled factory arrangement for the volume production of motorcars. Company policy was to scrap machines as fast as they could be replaced by improved types, and by 1912, the tool department was constantly devising specialized new machine tools that would increase output. By 1914, about fifteen thousand machines had been installed. Time and motion studies led in 1912 to the installation of continuous conveyor belts to bring materials to the assembly lines. By summer 1913, magnetos, motors, and transmissions were assembled on moving lines. After the production from these subassembly lines threatened to flood the final assembly line, a moving chassis assembly line was installed, reducing the time of chassis assembly from twelve hours and thirty minutes in October, to two hours and thirty minutes by December 30, 1913. The moving lines were at first pulled by rope and windlass, but on January 14, 1914, an endless chain was installed. "Every piece of work in the shop moves," boasted Ford in 1922. "It may move on hooks or overhead chains going to assembly in the exact order in which the parts are required; it may travel on a moving platform, or it may go by gravity, but the point is that there is no lifting or trucking of anything other than materials."

These mass production techniques greatly decreased unit production costs, which Ford passed on to the consumer in the form of lower Model T prices, to a low of only $290 for the coupe by 1927. Soon applied to the manufacture of many other items, these mass production techniques significantly increased the standard of living of the average American family and shifted the American economy from a production-oriented economy of scarcity to a consumption-oriented economy of affluence. Recognizing ahead of his contemporaries that mass production necessitated mass consumption and workers had to have the means to purchase what the machine

One of the first applications of the moving assembly line was this magneto assembly operation at Ford's Highland Park plant in 1913. Magnetos were pushed from one workman to the next, reducing production time by about one-half.
Source: Ford Motor Company

The body drop was the final step in assembling the Model T at Highland Park. Eventually Ford sped up the assembly process so that a Model T was completed every ten seconds.
Source: Ford Motor Company

produced, Ford inaugurated the five-dollar, eight-hour day for his workers in 1914, thereby more than doubling wages for a shorter workday. Mass production also created a new class of semiskilled workers, who needed neither substantial training nor physical strength. New remunerative employment opportunities were thus opened to the immigrant, black migrant to the northern city, women, physically handicapped, and educable mentally retarded. Unable to compete with Ford's progressive lowering of Model T prices through mass production, the makers of moderately priced cars in 1916 began to sell cars on credit, and by 1926, installment sales accounted for about two-thirds of all new car sales. Led by the automobile industry, the credit buying of expensive items became an established part of American life by the mid-1920s. Finally, the huge outlays of capital necessitated by mass production for gigantic specialized plants and complex machinery were beyond the means of the small automobile producer. Inevitably, automobile manufacturing changed from an industry characterized by intense competition among many small firms into an oligopoly by the late 1920s, with Ford, General Motors, and Chrysler accounting for about 80 percent of the industry's total output.

By the mid-1920s, automobile manufacturing ranked first in value of product and third in value of exports among American industries. The automobile industry was the lifeblood of the petroleum industry, one of the chief customers of the steel industry, and biggest consumer of many other industrial products, including plate glass, rubber, and lacquers. The technologies of these ancillary industries, particularly steel and petroleum, were revolutionized by the new demands of motorcar manufacturing. The construction of streets and highways was the second-largest item of governmental expenditure during the 1920s. The motorcar was responsible for a suburban real estate and construction boom (which has continued into the present, permitting most American families to live in single-family residences), and the rise of many new small businesses, such as service stations and tourist accommodations. With the new mobility of the population, business locations and residential patterns became more decentralized, and the Sunday drive to the country as well as the annual automobile vacation to distant national parks became commonplace. The Model T permitted farmers to shop in the city, killing off the crossroads general store and mail-order houses. Conditions of rural life and farm labor were also radically altered by the school bus, which replaced the one-room school with the consolidated and graded school, and the small farm tractor—exemplified by the Fordson—which diminished the need for farmworkers while greatly increasing agricultural productivity and alleviating much of the backbreaking drudgery of farm labor. (The larger economic units required for the efficient utilization of the tractor and the other mechanized equipment its use encouraged, however, ultimately made the small family farm obsolete.) Long-distance trucking, developed during World War I, opened up the Pacific Coast and Southwest to commercial exploitation, and knit regional economies more tightly together into a national economy. Noting the central role of automobility to changes in the American society and economy in the 1920s, the eminent historian Thomas C. Cochran has concluded, "No one has or perhaps can reliably estimate the vast size of capital invested in reshaping society to fit the automobile. ... This total capital investment was probably the major factor in the boom of the 1920's and hence in the glorification of American business."

By the end of the 1920s, the automobile had triumphed in the United States. The American society and economy had been restructured by the motor vehicle, and automobile ownership had been extended as far as the income distribution of the day would permit. The year that the Model T was withdrawn from production, 1927, was also the first year that replacement demand for new cars exceeded demand from first-time owners and multiple car purchasers combined. In 1929, the last year of the automobility-induced boom, the 26.7 million motor vehicles registered in the United States traveled an estimated 198 billion miles. During the Great Depression of the 1930s, although automobile sales slipped drastically, the miles traveled annually by motor vehicle in the United States continued to increase, and automobility proved to be one of the most depression-proof aspects of American life. Yet World War II brought a halt to passenger car production, and greatly curtailed automobile use through the rationing of gasoline and tires. Consequently, the automobile industry's record 1929 output of 5.3 million units was not again equaled until 1949.

With the triumph of the automobile in America came the end of the Model T era. As the Model T met the need for basic transportation and rural roads were vastly improved, consumers became more comfort and style conscious. Owners whose first car was a utilitarian Model T tended to trade up to larger, more powerful, and more stylish cars. Ford failed to keep abreast of this change in demand, and his competitors at General Motors and Chrysler not only surpassed him in product design but also by the mid-1920s caught up to Ford in the efficiency of their assembly lines. Except for minor face-lifts, and the incorporation of such basic improvements as the self-starter and closed body, the Model T remained unchanged long after it was outmoded. By 1927, secondhand cars of more expensive makes in excellent condition could be bought for about the same price as an obsolete Model T, and an annually restyled and far better equipped new Chevrolet cost only $200 more. In 1927, Chevrolet first surpassed Ford in sales. And by 1936, the Ford Motor Company had declined to third place in sales of passenger cars, with 22.44 percent of the market versus 43.12 percent for General Motors and 25.03 percent for Chrysler. The Ford Motor Company continued to decline until its aging founder turned over its presidency to his grandson, Henry Ford II, in 1945.

During the 1920s, the achievements of American capitalism were most prominently symbolized by Ford, who was idolized not only by the common people of the United States but also throughout the world. Ford received several thousand letters a day from ordinary people who were impressed with his accomplishments, and more was written about Ford during his lifetime (most of it adulatory) than any figure in American history. Even peasants in remote Asian villages knew about his Model T. And in the Soviet Union, the world's leading industrial capitalist was hailed as an economic innovator second only to Lenin for his contributions to mass production and the Fordson tractor.

Actually, as Reynold M. Wik, John Kenneth Galbraith, and others have shown, the success of the Model T was the result of a group effort by an outstanding entrepreneurial team at the Ford Motor Company, not the product of Ford's individual genius. And contrary to popular myth, Ford did not invent mass production; its constituent elements were all part of an evolving American manufacturing tradition, and many Ford employees contributed the key ideas that led to its implementation and perfection at the Ford Highland Park plant. After 1914, however, Ford Motor Company public relations policy was to promote Ford, not the company, and downplay the contributions of others to the success of the mass-produced Model T. The Ford mystique was thus in large part based on myths, which Ford deluded himself into believing and unfortunately encouraged.

By the time of his death on April 7, 1947, Ford's reputation had been tarnished by the blatantly anti-Semitic articles published in his *Dearborn*

Independent (for which he publicly apologized in 1927) as well as the deterioration of working conditions in his plants and his opposition to the unionization of his workers in the 1930s. Ford had bought out his minority stockholders in 1919, and after that tended increasingly toward arbitrary, one-man rule of his industrial empire. An increasing number of ex-Ford executives called this "Prussianization," and the *New York Times* in 1928 went so far as to call Ford "an industrial fascist." Ford refused to cooperate with President Franklin Delano Roosevelt's plans for recovery from the Depression. Strokes suffered in 1938 and 1940 left Ford's eroding mental faculties greatly impaired.

Nevertheless, Ford remains a folk hero to millions of people throughout the world, who, ignoring his failings, continue to identify him as the progenitor of mass personal automobility and mass production. Ford will probably remain an important figure in world history long after the names of most American presidents are forgotten.

18   A. C. Gilbert, Toys, and the Boy Engineer

Carroll Pursell

When A. C. Gilbert died in 1961, his *New York Times* obituary referred to his having "blended the familiar qualities of Frank Merriwell, Theodore Roosevelt, Peter Pan and Horatio Alger heroes." During his long career, he had an enthusiastic, almost-boisterous commitment to the idea of encouraging boys (but not girls) to explore as well as enjoy science and engineering. It was a project very much in tune with the traditional gender norms of the period. Its "modern" commitment to the benevolence of technological "progress" was coupled with a belief that American children were "naturally" inclined toward science and technology. That link would ensure the nation's safe and prosperous future. By the time of his death, Gilbert was credited with some 150 patents, mostly for toys and their components, notably the famous construction kit, Erector. On the foundation of these patents along with the strategic acquisition of other companies, he built a manufacturing enterprise that helped shape the toy industry during the first half of the twentieth century.

Alfred Carlton Gilbert was born in Salem, Oregon, on February 15, 1884. His father was a banker, and Gilbert grew up comfortably in what were sometimes frontier conditions. He was drawn to competitive sports at an early age and claimed to have won a tricycle race at the age of seven. For most of his life, he found athletic competition to be a compelling outlet for his energy and drive to excel. He told of building a gym in the family barn, when living for a few years in Moscow, Idaho, and fitting it out with various pieces of equipment, such as importantly, early on, a punching bag that he had won by selling subscriptions of the magazine *Youth's Companion*. His prowess with the bag led to his running away from home while still in his early teens, joining a traveling minstrel show, and being featured as "THE CHAMPION BOY BAG-PUNCHER OF THE WORLD!" It was a short career as his father soon caught up with him and brought him home. While living in Moscow, he also helped organize the Moscow Athletic Club among the other boys, and together they held track and field meets. At the nearby University of Idaho, he saw his first pole-vaulting competition and became a lifelong enthusiast of the sport.

When Gilbert's older brother Harold moved to Forest Grove, Oregon, to attend Pacific University, he went along and enrolled in the Tualatin Academy, the university's preparatory school. While doing well enough in his academic studies, he seems to have devoted a good deal of time and effort to what he called "tomfoolery" and athletics. He broke the "world chinning record" by pulling himself up forty consecutive times. More important, he was allowed to compete on the university's track and field team, where he experienced his first professional coaching. He decided to become a physical education teacher himself and enrolled for two summers at the School of Physical Education in Chautauqua, New York. A number of the instructors he met were medical doctors, including several who had attended Yale University. Finishing his time at Pacific University, he went east to earn a medical degree and athletic renown at Yale.

Gilbert found his medical studies hard work, but he made time for a successful athletic career, concentrating on the pole vault. He had set records in the event while at Pacific, but at Yale he went all the way to the London Olympics in 1908, setting a world's record in the Olympic tryouts and winning a gold medal at the games themselves. Part of his success was his adoption of two technical innovations: the bamboo pole, and the use of a hole in the ground, rather than a metal point on the pole, to launch himself into the air. At both Pacific and Yale, he participated in a number of sports, and in his autobiography, an appended list of "Notable Events in the Life of A. C. Gilbert" begins with twelve sporting successes between 1901 and 1909. The thirteenth on the list is the 1913 introduction of the Erector set.

It was also at Yale that he started to stage magic shows. Besides the punching bag, his magazine sales in Salem had won him a kit of magic tricks and an instruction book. At Yale he often entertained his friends with his tricks and occasionally earned some money for entertaining at children's parties. He graduated to putting on shows at clubs from Boston to New York, raising his fees. When he returned to New Haven, Connecticut, after the Olympics, he and a friend began to produce boxes of magic

An advertisement for Erector.
Source: *Parent's Magazine* 8 (December 1933): 2

tricks that they sold through a mail-order catalog. He devised some of the apparatus, and his friend, John Petrie, the rest. They called their enterprise the Mysto Manufacturing Company, and produced three different boxes of cards, coins, rings, and "vanishing apparatus," which sold for twenty-five cents, fifty cents, and a dollar, respectively. Though Gilbert graduated with a medical degree, he turned his full attention to building his Mysto Manufacturing Company. His father lent him five thousand dollars so that he could acquire a manufacturing facility. In 1909, he and Petrie built their two-story wooden factory. They opened their own magic store in New York in 1910, and incorporated the next year, going on to open stores in Philadelphia and Chicago.

On a sales trip to the West Coast, Gilbert had found that he was dealing mainly with the toy buyers in the many stores he visited. He gained the impression, he later wrote, that "there weren't many good toys and that most of them were made in Germany." On his visits to the New York store, he looked out the train window at girders being erected to carry power lines as the railroad switched from steam to electricity. Gilbert found the girder work fascinating and thought that most boys would too, so he linked all this with his growing interest in the toy business. The idea for the Erector set came to him, he explained, "all of a sudden" in fall 1911.

This origin myth was remarkably like that of an English construction toy introduced ten years earlier by the inventor Frank Hornby. According to Hornby, while traveling by train from London, he saw out the window a crane lifting material in a goods yard. He realized that he could make a model crane out of strips of metal. Hornby further realized, he recalled, that if he went to all the trouble of making the many parts for the crane, it would be best to standardize them so that he could use them to make other machines as well. He patented his idea for the construction kit, calling it Mechanics Made Easy, but later changing the name to Meccano. It was sold in the United States and was, for many years, Erector's closest competitor.

Immediately on experiencing his epiphany, Gilbert began cutting patterns for parts out of cardboard. He then had a worker at his factory make a set out of steel, and started fastening them together with nuts and bolts. Gilbert found that in order to add rigidity to long strips when bolted end to end, he had to design a groove along the edge of each piece. He considered this "the most important single factor in the success of Erector." He also included rods, pulleys, and eventually an electric motor to aid the toy's principles of "action and motion." He was enthusiastic about the toy's potential and astonished at his associates' lack of interest. Gilbert finally decided that they were solely interested in magic, and with help from his father, was able to buy them out.

Before the end of 1911, Gilbert acquired a former carriage works in New Haven, where he began to assemble the machinery to manufacture Erector. Like Hornby, he was faced with mass-producing a range of identical parts and required special machines, many of which were designed or modified in the factory itself. One, standing twenty feet high and extending half a block, was used for zinc plating the millions of Erector parts. When the company became heavily committed to small fractional electric motors for Erector and other products, it made its own, developing a novel method of using enamel insulation for the miles of wire used to wind armatures. This last was the invention of one of his workers, John Lanz, but Gilbert took great pride (and always a share of the credit) in the special machinery and processes they developed. The special machinery, he wrote, was "the part of the business that fascinates me."

While the magic business continued to grow, Gilbert's concentration was on rolling out Erector. He displayed it first at the 1913 toy fair in New York. As he described Erector, "it had wheels, axels, gears, and pinions in addition to its girder parts of different sizes, its angles and plates. Most important, it had an electric motor." Its success was repeated at a toy fair in Chicago, and the company had to expand its production to keep up with the many new orders. Gilbert spent a great deal of time designing models that could be made with Erector, testing them to be sure they actually worked, and then adding them to the instruction book that accompanied each set. He hoped and believed that boys who had the set would create their own designs, but might need help getting started.

Again like Hornby, Gilbert decided to undertake an extensive advertising campaign for Erector, with large ads in magazines with a national circulation. The ads began to appear in 1913, and most of them carried the headline "Hello, Boys! Make Lots of Toys." He also started a small newspaper called *Erector Tips*, which was an important vehicle for promoting contests, with cash prizes, for the best new models invented by customers. By 1933, the prizes had grown: second prize was a new Chevrolet, and first was a trip to "the Panama Canal, or Boulder Dam or the Empire State Building or any other engineering project in the United States." Gilbert had also added a Sunday evening radio program, "Engineering Thrills," featuring "true stories about real engineers and their hair-raising adventures in digging the Panama Canal, building bridges and skyscrapers."

The emphasis on boys was both deliberate and unapologetic. Gilbert always insisted that with Erector, boys did real engineering, a professional activity then as now overwhelmingly male dominated. Furthermore, Erector developed character and skills beyond mere technical knowledge. In the introduction to his 1920 catalog, he wrote,

I feel that every boy should be trained for leadership. It is only the bright-eyed, red-blooded boy who has learned things, done things, dared things beyond the reach of most boys who will find the way open to really big achievements. ... My toys are toys for the live-wire boy, who likes lots of fun and at the same time wants to do some of the big engineering things—things that are real—things that are genuine.

One ad in 1933 featured a boy playing with his Erector set, but had an inset message starting with the familiar "Hello Boys!" and then showing an image of Gilbert's face followed by the lines "you build like a real engineer when you build with ERECTOR." Gilbert always claimed that many scientists and engineers had gotten their start by playing with Erector or his chemistry set.

Over the years Gilbert added other toys to his line, such as a machine gun offered in 1920. He enthused that "if there ever was a real live-wire toy for the red-blooded boy, this is it. Say, you can have more genuine sport with this machine gun than anything I know of. It's the real thing." Another wartime spin-off was a nurse's kit, one of the few toys that Gilbert offered for girls. Introducing his toy tractor, he noted, "Every wide-awake boy knows what wonders the Tractor has accomplished, and what a tremendous aid it is in the great farming districts of the West. You boys want to see how these up-to-the-minute machines work." In 1938, Gilbert bought out the American Flyer Corporation, and after making numerous improvements, reintroduced the electric train system that became a major offering of the company. Eventually the American Flyer trains and scale-model HO sets he also developed were on display with the other toys at the Gilbert Hall of Science on Madison Square in New York City. In 1950, he debuted the television program *Boys' Railroad Club*.

Gilbert began work in 1915 on what he envisioned as a complete line of educational toys, including an obligatory chemistry set. Five years later, the company introduced a number of new kits, or sets, covering different fields of scientific and engineering activity. The 1920 catalog titled *Gilbert Boy Engineering* listed the sets on its cover: Civil Engineering, Weather Bureau, Hydraulic and Pneumatic Engineering, Light Experiments, Radio Engineering, Electrical Engineering, Constructional Engineering, Chemical Engineering, Signal Engineering, and Carpentry.

One of the last such kits introduced by the company was an Atomic Energy Laboratory, brought out in 1950. Gilbert claimed that he had the encouragement of the government and had worked closely with MIT in its development. It was, he wrote, "the most thoroughly scientific toy we had ever produced." It was expensive at $50 a set, but contained, besides some radioactive materials, a Geiger-Mueller counter, cloud chamber, spinthari-scope (a device to observe individual nuclear disintegrations), and electro-scope to measure the radioactivity of different materials. It was mocked as "the world's most dangerous toy" and proved to be less than successful on the market, in part because of its high price, and in part because it was too sophisticated for the typical boys who made up its customer base.

Because 90 percent of its toys were sold at Christmastime, the Gilbert Company had trouble keeping its plant and employees busy year-round.

An advertisement for Gilbert's Atomic Energy Lab.
Source: Wikimedia Commons

One solution, since it manufactured large numbers of small electric motors for its toys, was to branch out into small home appliances that could also be motor driven. In 1918, Gilbert filed a patent for an electric fan, the novelty of which was, he claimed, the guard that surrounded the blades. He was granted a patent in 1933 for an electric drill, "a relatively light, portable electric tool which may be supported in the hand of an operator, and which, while efficient in use, is of relatively simple form and may be cheaply made." In 1928, he received a patent for a handheld vibrator, "such as frequently used by barbers in giving face massages, or by physicians and others in the treatment of various diseases, and for other purposes." It was the "most significant of our new products," he observed, because it was "the beginning of a series of important motor-driven appliances that were

April 25, 1933.   A. C. GILBERT   1,905,462

ELECTRIC DRILL OR SIMILAR TOOL

Filed May 20, 1929

Fig. 1.

Fig. 2.

Fig. 4.

Fig. 3.

Inventor

Alfred C. Gilbert

By Rockwell & Bartholow

Attorneys

The patent for "Electric Drill or Similar Tool," issued to A. C. Gilbert.

lightweight and attractive and that could be used by a woman in the home. It was to become one of the steadiest sellers in the appliance field."

During his active business years, Gilbert was a leader in the toy industry. In 1916, he was one of the forty-six founders of the Toy Manufacturers of the U.S.A., later the Toy Industry Association, and became its first president. During World War I, he led a delegation of industry leaders to lobby in Washington, D.C., for the continued production of toys even during the wartime period of restricted access to vital materials—hence his nickname, "The Man Who Saved Christmas," also the name of a television film in 2002 starring Jason Alexander as Gilbert. By the mid-1930s, the company had sold more than thirty million Erector sets and was the largest employer in New Haven, with over five thousand workers. Gilbert was one of the seven original inductees to the Toy Industry Hall of Fame in 1985. He retired as company head in 1954, and died seven years later in 1961. After 1954, his son A. C. Jr. ran the company until it was sold in 1962. The company ceased production in 1966, and was declared bankrupt in 1967. Ironically, the rights to the Erector set were bought by the Mecanno Company, which continued its manufacture. The fate of the A. C. Gilbert Company followed a pattern familiar from other successful firms founded and dominated by a strong, charismatic figure. As the historian Gary Cross has written, "The heart of Gilbert's enterprise was his own highly publicized personality." When he died, so did his firm.

Gilbert's hearty "Hello Boys" greeting in his advertisements was a gender bias that grew out of his own commitment to strenuous masculinity; as he once wrote, "I've remained a boy at heart." But the same bias was rampant in the toy industry as a whole. Then as now, toys directed at boys were more numerous, more active, and more complex than those for girls. Boys were urged to investigate, experiment, and innovate—activities designed to develop each boy up to his highest capacity. Girls were not so encouraged, and were given dolls, small household tools, and appliances that prepared them not for success in the modern world of scientific and technological change but rather the more static domestic occupations of their mothers. No better example could be found than the offerings of the LaVelle Toy Company, a Gilbert spin-off established in the 1920s but kept completely separate from the Gilbert enterprise; it is completely absent, for instance, from his autobiography, *The Man Who Lives in Paradise*. During its brief existence, LaVelle kept its offerings carefully within the gender expectations of the time, such as a Little Laundress kit, millinery set, and Nurse's Outfit. This last was simply advertised as "Joy and Fun for Little Ladies."

In her study of the close cultural ties between science and childhood in twentieth-century America, Rebecca Onion makes the point that while children were declared to be little scientists—curious, experimental, and eager to learn about the rules and makeup of the natural world—the belief was widely limited to boys. This was consistent with the cultural association of both science and engineering with masculine values as well as male practitioners. It also shaped the toy world, even as that helped shape the world of boys.

Gilbert did not invent the trope of the boy engineer, but he used it brilliantly to his company's advantage. Especially in the years since World War II, however, the government, technical schools, professional organizations, and others have made efforts to lure more girls into science and especially engineering. Yet female participation has changed little in the last half century, and the barriers of masculine privilege have held firm. Just as Gilbert and his industrial colleagues had deliberately tried to prepare boys for technical careers, so in recent time have a few toy manufacturers sought to attract girls to science, technology, engineering, and mathematics subjects. Such efforts, however, must overcome a long tradition of cultural resistance.

19   Peter L. Jensen and the Amplification of Sound

W. David Lewis

During the relatively short span of less than two hundred years, human communication has been transformed by a series of technological innovations including high-speed presses and typesetting machines, the telegraph, the telephone, the phonograph, radio, television, the digital computer, and orbiting earth satellites. Of key importance in the unfolding of this process has been the growth of knowledge about electricity, which had begun to yield a flood of practical applications by the last quarter of the nineteenth century. As scientists and engineers explored the behavior of alternating current and electromagnetic waves, and learned about the presence of electrons in the basic structure of the atom, the pace of change accelerated in the early twentieth century; the exciting new technology of radio served as a major focus of activity. Among the many inventions that resulted was the creation of a new and unheralded communications device: the loudspeaker.

Electronically amplified sound is such an omnipresent feature of modem society—sometimes reaching levels of volume so intense that it constitutes an environmental hazard—that it is difficult to envision a world in which it did not exist. Yet it was only in 1915 that Peter Laurits Jensen, a Danish immigrant, and Edwin S. Pridham, his American associate, invented the modern loudspeaker in a small bungalow on the outskirts of Napa, California. As has been true of many other inventions, its birth was accidental. Jensen and Pridham had been working on an improved telephone receiver, but could not reduce its cost and size to the point at which it could be commercially exploited. Acting on a suggestion by a relative of Jensen's wife, the two inventors transformed their brainchild into a device that for the first time in history, electrically amplified sound far beyond any previous limits.

The invention of the loudspeaker is therefore a classic example of serendipity, the art of finding things unintentionally or by indirection. More than this, however, it illustrates the complex process of feedback that took place as accumulating scientific and engineering knowledge produced rapid changes in radio technology in the early twentieth century. The loudspeaker's technology represented a shift away from what historian Hugh G. J. Aitken has called "syntony and spark," involving methods of transmission and reception that characterized early radio, to a "continuous-wave" technology that opened up dramatic new possibilities after the turn of the century and laid the foundations of radio as we know it today.

Jensen's role in the development of the loudspeaker resulted from his employment as a young man in his native Denmark under one of the pioneers of continuous-wave technology, Valdemar Poulsen. The opportunity to come to America to help install continuous-wave equipment for entrepreneurs who had acquired the right to exploit Poulsen's patent rights in the United States launched a remarkable career that demonstrates the process of "technology transfer" from one part of the world to another within scientific, technological, and business communities whose knowledge and activities transcended national boundaries. In time the young immigrant would help lay the foundations for modern audio technology, and win both financial success and international recognition. The reputation gained by such products as Magnavox radio and phonograph sets as well as Jensen speakers testified to the effectiveness and importance of his work.

Jensen's career, though, was not dedicated solely to technological innovation. A sensitive and reflective person, he was obliged by the circumstances of his life to search for a new sense of personal identity as a stranger in a land that however much it differed from the country of his birth, he came to love. Absorbing himself in its history and traditions through the influence of a fellow engineer who became his best friend, he sought diligently to make himself over into a new self-image only to experience in a moment of profound introspection that he belonged to two countries and would always bear the stamp of both. His life belies a common image of engineers as persons who lack interest in emotional or cultural growth. Clearly he was a complex man of much substance and inner depth.

Nothing in Jensen's parentage or background suggested that he would become an inventor and make significant contributions to the development of electronic technology. Born May 16, 1886, near Stubbekobing on the

Peter L. Jensen, circa 1930.

island of Falster in the eastern archipelago of Denmark, he seemed destined to follow the career of his father, Lods Ole Jensen, a navigator who guided vessels through the waters between Falster and the neighboring island of Lolland. As Peter later stated in his autobiography, acquaintance with the Baltic Sea, "sometimes calm and innocent looking, but often deadly in its fury," began early. At age seven, he started assisting his father at the rudder of a small pilot boat that plied the turbulent Strait of Groensund while a larger sailing ship came on behind, "splashing and foaming at the bow," and sometimes threatening to collide with its tiny escort in the treacherous passage. At other times he fished for herring, or helped his mother, Hansine Petersen Jensen, pick potatoes, weed beets, or tend livestock on the small plot of agricultural land that surrounded the family home. The parenting he received was austere; although he was certain of his mother's love, he could later remember no occasion on which she had ever shown him any physical affection.

Escape from this life of danger and drudgery was made possible by the precocity that Jensen displayed in his elementary schooling at the nearby village of Moseby, where he ranked at the top of his class. When he was thirteen, the senior teacher paid an unexpected visit to his parents. "No child had ever gone from his to a higher school to graduate from the state university," Jensen later wrote, "and he came to suggest that I do just that." After much debate his parents assented, and Jensen left home for a boarding school in Norre Alslev, passing the entrance examination for the University of Copenhagen three years later. He never matriculated, however, for his father died from the effects of "one of the worst storms ever to sweep down on the Baltic," forcing Peter to return home to look after his mother. Taking a job hauling fagots from a woodlot to a sawmill, his knowledge of German and English attracted the attention of a superintendent, who then suggested that he ought to use his education to better advantage and urged his mother to send him to Copenhagen in search of opportunity. She did so, giving him a letter to a well-known resident of that city, Lemvig Fog, whose family had at one time boarded at the Jensen home during a summer vacation.

Fog was an engineer-entrepreneur who had returned to Denmark after making a fortune in Brazil. As a financial backer of Poulsen, he suggested that Jensen embark on a technical career and helped him secure employment as a "mechanist apprentice" in Poulsen's laboratory. Here Jensen received training comparable to that obtained by American counterparts in what historian Monte Calvert has called the "shop culture" approach to engineering education in the United States—at that time, still a viable route to a professional career despite an increasing trend toward study in an institute of technology or university engineering department. Recognizing that both Fog and Poulsen "were known as engineers and yet neither had graduated from a university," Jensen resolved to follow in their footsteps.

Poulsen was already an inventor of international consequence when Jensen entered his laboratory; at the Paris Universal Exhibition of 1900, he had won a grand prize for a device called the "Telegraphone," an early forerunner of the tape recorder. Shortly after Jansen began working for him, Poulsen won further recognition when he succeeded in amplifying electromagnetic waves generated by an arc transmitter by having the arc burn in an atmosphere of hydrocarbon vapor instead of air. A significant advance in the development of continuous-wave technology, the "Poulsen arc" was a finicky device that permitted sharp tuning but required frequent regulation in order to work efficiently. Jensen became adept at keeping it in effective operation and was entrusted with this responsibility by Poulsen.

In 1905, Poulsen set up a radio station in Lyngby, near Copenhagen. Jensen, who had taught himself Morse code, and was adept at sending and receiving radio messages, took part in its construction. Efforts to capitalize on the transmitting capabilities of the Poulsen arc, though, were impeded by its limited range and the deficiencies of the Branly coherer, a glass tube within which metal filings between two electrodes were excited by damped waves generated by a spark transmitter, thus constituting the standard signal detector in the early days of radio. With the help of P. O. Pedersen, a professor in the Royal Technical College, Poulsen developed a new device known as the "tikker," which used a vibrating reed, two gold wires acting as an interrupter, and telephone circuitry enabling radio signals to be received through a pair of headphones, or in a high-speed version, to be registered on photographic tape for subsequent scanning. Using a crystal detector, efforts were made to transmit human speech with this equipment—a feat that was impossible with spark technology. While taking part in this project, Jensen seized on the idea of placing a microphone in the ground circuit of the transmitter and connecting the crystal detector to a grounded tikker as a receiver, with successful results. Before serving a brief stint as a wireless operator in the Danish Navy in 1909, Jensen utilized the new system by transmitting recorded music from Lyngby to a small number of listeners, including radio operators on ships at sea.

Although Poulsen's system was less expensive and easier to switch from one frequency to another than an alternate continuous-wave system developed by Reginald Fessenden in the United States, it saw only experimental use until it attracted the attention of a group of American investors represented by Cyril F. Elwell, a native of Australia who had emigrated to California and earned a degree in electrical engineering at Stanford University. Elwell had become interested in radio as a result of testing an unsuccessful system of wireless telephony developed by Francis J. McCarthy, a young inventor whose work had aroused interest among San Francisco businessmen prior to his death in a 1906 automobile accident. Seeing in Poulsen's system a potential rival to the one that Fessenden had developed in cooperation with General Electric, Elwell went to Copenhagen in 1908 and was impressed by his observations of the Danish technology, although somewhat deterred by the stiff terms demanded by Poulsen and his associates for its exploitation in America. Returning in 1909 with backing from a group of investors in Palo Alto, California, Elwell persuaded the Poulsen syndicate to modify its demands and take part in a California enterprise to be known as the Poulsen Wireless Telephone and Telegraph Company (PWTTC).

When Poulsen asked him to go to America to supervise the installation of the Danish equipment, Jensen accepted the offer as the opportunity of a lifetime. On December 9, 1909, he left Denmark with a mechanic, Carl Albertus, who had previously lived in Chicago. Arriving in New York City after a twelve-day voyage, the two young men traveled across the continent by rail to San Francisco. Jensen marveled at the contrast between America, where he "found a most peculiar conglomeration of the best and the worst," and his native country, where everything was "clean and orderly with no extremes." The "semi-tropical brightness of California made the deepest impression upon me," he later recalled. "I had never seen palms and tropical fruit trees before, and the weather as we rolled down into the Sacramento Valley was sparkingly bright." Crossing the Bay from Oakland into San Francisco at sundown evoked a feeling that he "had arrived at the end of the world and found that end most glorious." Equipment soon arrived from Denmark, and two stations were erected in the San Joaquin Valley, one in Sacramento and the other about fifty miles away in Stockton.

During completion of the Sacramento radio station, Jensen met Pridham, a native of Illinois who had graduated from Stanford in 1908 with a degree in electrical engineering after working his way through school by cleaning bricks in debris left over from the 1906 San Francisco earthquake. Fascinated by radio, he had taken a job with the PWTTC shortly after graduating. Pridham became Jensen's "ideal of a real American." He helped Jensen polish the rudimentary English he had learned in Denmark, and taught him the lore and traditions of the United States so successfully that the study of American history became Jensen's hobby. "During the next fifteen years," Jensen later wrote of Pridham, "he never ceased in his efforts to change me from a foreigner into a true American."

Becoming inseparable friends, Jensen and Pridham soon moved to San Francisco, where a third radio station was built on the beach with two tall masts that became landmarks to local residents. Jensen later marveled that he and Pridham were oblivious to the commercial possibilities of radio broadcasting, despite the fact that amateur operators with crystal sets listened to transmissions from the new station to Sacramento and Stockton. Like the PWTTC's backers, they had little sense of radio's market possibilities, seeing it mainly as a potential competitor of the telegraph as well as a means of ship-to-ship and ship-to-shore communication. Dominated by the Marconi Wireless Telegraph Company of America, the radio industry in the United States was characterized almost exclusively by the use of spark technology to send messages in Morse code. Although such inventors as Fessenden, John S. Stone, and Lee de Forest were developing significant new devices utilizing continuous-wave technology for transmission, tuning, and reception, they lacked effective strategies for competing successfully with the Marconi interests, and the value of their work was unappreciated by the general public. De Forest in particular had a version of how "wireless" might be used to broadcast news, advertising, and music, and had invented a three-element tube known as the "audion" that along with powerful alternators developed by Fessenden, would in time make spark technology obsolete. For the present, though, only the activities of amateur operators gave any hint of the future of broadcasting; the use of radio was otherwise confined to a relatively few business institutions and governmental agencies, particularly the U.S. Navy.

As additional stations were built by Poulsen's American associates in Los Angeles and other locations, the PWTTC went through a process of reorganization leading to the establishment of two new firms: the Poulsen Wireless Corporation, and its operating subsidiary, the Federal Telegraph Company (FTC). Considering his work done, Jensen decided to go back to Denmark. Before he could leave, however, John C. Coburn, a glib promoter who had been active in selling stock for the PWTTC, approached him. Having unloaded his investment in that company at a substantial profit, Coburn now represented a group of San Francisco businessmen who had failed to secure stock in the FTC and were engaged in a grandiose scheme to buy the rights to exploit Poulsen's patents throughout the entire world out-side the United States, hoping to acquire the remaining American rights at a later time. Believing that Jensen's influence would be useful in selling the idea to Poulsen, Coburn promised to make the young Dane chief engineer of the resulting enterprise and give him a large block of its stock.

Dubious about the project's chances for success, Jensen nevertheless agreed to go to Copenhagen on behalf of its promoters, stipulating only that Pridham must be allowed to go along. Resigning from the FTC, the two friends went by way of England with the hope of getting permission to erect stations in Ireland and Canada, only to find that Poulsen had already granted an option on his English patent rights to a British syndi-cate. Proceeding to Copenhagen, they were politely received by Poulsen, who pleased Jensen by now addressing him as "Mr." in acknowledgment of the enhanced status he had gained through his experience in America. Yet the Danish inventor refused to participate in the scheme proposed by the San Francisco syndicate.

By this time, perhaps because of Pridham's influence, Jensen had decided to return to the United States and become an American citizen. After visit-ing his mother, he and Pridham headed back by way of London, sailing from there to New York City on the *Lusitania*. As the ship came through the Narrows and Jensen gazed at the Statue of Liberty, a "new and strong feeling of patriotism" swept over him. "I knew this country would be mine, forever," he later declared.

Returning to California, Jensen and Pridham were introduced by Coburn to Richard O'Connor, a San Francisco soap and candle manufacturer with powerful connections in local and state politics. Impressing Jensen as "the ideal type of politician in the best sense of the word," O'Connor was a strong backer of Hiram Johnson, who became governor of California in 1910 on a progressive reform ticket. O'Connor and his business associates had been involved in the abortive plan to acquire Poulsen's world patent rights, and were still eager to invest in radio despite the failure of the recent mission to Copenhagen. With their support, an enterprise called the Commercial Wireless and Development Company (CWDC) was organized, with O'Connor as president and treasurer. As the firm's engineers, Jensen and Pridham were to conduct research and experimentation in radio, and take out patents on anything they might invent. On February 22, 1911, they moved to Napa, about thirty-five miles northwest of San Francisco, choosing the community for its relative seclusion. Acquiring "a small bungalow and a good sized lot," they assembled some equipment and commenced work. Within a short time, Jensen met and married a Californian, Vivian Steves. By 1914, they had a daughter, Jean, and son, Karl. Two more children, Patricia and Marian, followed in, respectively, 1917 and 1925.

The project in which Jensen and Pridham became engaged in their Napa laboratory epitomizes the process of technological transfer that took place between key industries—in this case involving the telegraph, telephone, and radio—as the development of continuous-wave technology gained momentum. Joined by Albertus, who signed on as a mechanic, Jensen and Pridham became intrigued by the photographic apparatus that Poulsen had devised for recording high-speed transmissions of dots and dashes from his tikker. Poulsen had tried to sell rights to this device to Elwell and his American backers in 1908, but they had not been impressed by it. Struck by the rapidity with which the thin wire in the tikker responded to signals as it moved back and forth between the poles of a magnet, Jensen and Pridham thought it might respond just as rapidly to sound waves and furnish the basis for a telephone receiver. Substituting a somewhat-heavier wire for the thin one, they connected it to a diaphragm with a matchstick and integrated it into a telephone circuit with a microphone speaker. Testing the new apparatus, they were gratified to hear that the receiver reproduced human speech "with exceptional strength and clarity."

After making a careful search of the scientific and technical literature at their disposal, Jensen and Pridham mistakenly concluded that they had stumbled on a new method of sound reproduction. Applying for a patent covering what they described as the "electro-dynamic principle," they improved their new invention by suspending a coil of copper wire between stronger electromagnets, and experimenting with a variety of diaphragms and sound boxes. Believing that they had invented a telephone that performed better than any then in use, they received a setback when their patent application was denied; the principle they had thought to be original had in fact been anticipated by a number of scientists and inventors in Germany, Great Britain, and the United States. Because nothing had ever been done to exploit it commercially, they did manage to win a subsequent patent protecting only their specific device. Nevertheless, it was still too heavy and expensive to compete with existing telephone receivers, and they were unable to interest the American Telephone and Telegraph Company (AT&T) in purchasing the rights to it despite a trip they made to New York to talk to officials at AT&T's Western Electric Laboratories.

Bitterly disappointed, Jensen and Pridham talked about ending their partnership on the way back to California. Having invested approximately $30,000 in their work, most of the firm's backers were also ready to pull out, but O'Connor was unwavering in his support, and the venture stayed precariously alive. At this critical moment, Ray Galbreath, an uncle of Vivian Jensen, visited the Napa laboratory and suggested a new approach. Referring to a colorful local figure who made announcements with a megaphone at baseball games in the San Francisco area, Galbreath proposed that the failed receiver might serve as a new means of making public announcements. "If you can make it talk a little louder and put a horn on it, like Foghorn Murphy's," he said, "and if you put enough of them around a ball park, maybe we can understand what is being said a little better, and maybe they don't need Foghorn Murphy any more." This was a far cry from the lucrative telephone markets Jensen and Pridham had in mind, but there was also the prospect that the type of device Galbreath had suggested could be used in railroad stations to announce train departures and arrivals. Not knowing what else to do, they decided to proceed.

Taking a gooseneck horn from an old Edison phonograph, Jensen and Pridham designed a fitting to connect it with their receiver. Clustering six powerful microphones acquired during their European trip in 1910, they improvised an improved transmitter. To their original circuitry they added a transformer and twelve-volt storage battery. Without realizing it, they had created a sound system with a potential output approaching twenty-five watts, thousands of times greater than anything in existence. As Jensen later recalled, connecting the components resulted in "a screaming howling noise which was ear-splitting and terrifying," but they recognized this as a feedback phenomenon that could be corrected if the transmitter and receiver were placed far enough apart. Running a line to the roof of the bungalow, they attached the horn and receiver to the chimney, pointing northwest toward open countryside. When Pridham spoke through the microphone cluster inside the house, it sounded to Jensen "like a voice not of this earth. Had I closed my eyes it would have been easy to imagine that a supernatural colossus was shouting up the chimney."

Sprinting across the open fields with Albertus struggling unsuccessfully to keep up, Jensen found that by listening intently, he could understand what Pridham was saying a mile away. When he returned, Pridham bicycled to the same spot and listened for himself, after which the excited associates called O'Connor in San Francisco. Incredulous at the news, O'Connor insisted that it be repeated several times. Satisfied at last that it was true, he brought a group of stockholders to Napa the following day, accompanied by a lawyer. The successful demonstration that followed produced assurances "that finances would no longer be a problem." Casting about for a name to give their brainchild, Jensen and Pridham considered such alternatives as "Stentor" and "Telemegaphone," but finally settled on "Magnavox," a term meaning "great voice" in Latin. Ultimately it became universally known as the loudspeaker—a name Jensen had already considered but rejected because he found it unappealing.

At this point, Jensen and Pridham had gone through a process that historians of technology would subsequently call "invention," in which a new principle or device appears in rudimentary form. They now entered on a stage that scholars would later characterize as "development," in which modifications are made to improve a new technology to prepare it for commercial application in a final stage known as "innovation." Through 1915, Jensen and Pridham gradually improved their invention. Some of the changes they made merely affected its appearance, giving it the distinctive shape that would become characteristic of loudspeakers in the future. Others were more substantive, such as substituting new American-made microphone buttons for the previous European ones in order to enhance transmission. In the process, they created the prototype of the electric phonograph. Placing the loudspeaker in a cabinet, they mounted a turntable and microphone on top, and designed a volume control that could be turned up or down by manipulating a knob. A patent was quickly obtained, the only remotely comparable device being a box within which a telephone receiver served as a burglar alarm by emitting a whistling sound.

During the course of their work, Jensen and Pridham treated the townspeople of Napa to occasional concerts of recorded music from their rooftop. A local resident was startled one evening to hear an "extraordinary voice" coming from three-quarters of a mile away; it turned out to be that of the famous contralto and mezzo-soprano Ernestine Schumann-Heink, singing Christmas music. Ultimately the sound of the loudspeaker could be heard within a radius of seven miles.

Fearing that premature exposure of their system would stimulate efforts to copy it, Jensen and Pridham refrained from exhibiting it at the 1915 Panama Pacific World's Exposition in San Francisco. It was obvious that a formal public demonstration could not be indefinitely postponed, however, and on December 10, 1915, such an event took place before an invited group in the stadium at Golden Gate Park. Conditions were far from ideal; the wind was blowing at twenty miles per hour, there was a drizzle of rain, and some schoolboys played football on the field as the test took place. Nevertheless, reporter Edgar "Scoop" Gleason published an enthusiastic article about the demonstration the following day in the San Francisco *Bulletin*, commenting in glowing terms about "the slender tone of a single violin heard one mile away," the sound of operatic soprano Luisa Tetrazzini's voice reverberating throughout the stadium, and "a piano solo resembling the chimes of Westminster Abbey played by the Colossus of Rhodes." Although the loudspeaker was placed at a considerable distance from the listeners, the noise of the football game was inaudible as the music of a Hawaiian string quartet resounded "like the harps of the cyclops."

Two weeks later, on Christmas Eve, a much more elaborate airing of the system took place when a crowd estimated at a hundred thousand people gathered in the plaza in front of San Francisco's new city hall and listened to recorded music that sounded as if it had been "uttered by a giant" as part of an annual holiday celebration sponsored by the *Bulletin*. Still concerned about keeping the nature of their apparatus as secret as possible, Jensen and Pridham operated it from a balcony, the loudspeaker being concealed behind an American flag.

On December 30, the first indoor demonstration of the loudspeaker took place in San Francisco's new Civic Auditorium, a twelve-thousand-seat facility that had recently been given to the city by the Panama Pacific

Edwin S. Pridham (left) and Jensen in their laboratory in Napa, California, in 1915.

Exposition Company. A speech by Governor Johnson was to be transmitted to the hall from his mansion on Green Street, two miles away, and special wiring arrangements were made to facilitate this. Jensen, stationed in the hall while Pridham supervised transmitting equipment located in the mansion, was understandably nervous as the opening presentations began. He agonized as the architect who had designed the building, unaccustomed to the use of a loudspeaker, became alternately inaudible and overpowering as he waved the microphone back and forth while pointing to various features of the auditorium. After this ordeal was over, the young Dane precipitated a minor scuffle trying to disperse a choir that had massed in front of the horn from which the governor's voice was to emanate. Despite such problems, and the fact that Johnson's speech could be heard much better in the galleries than was possible on the main floor because of the placement of the choir, the address received a loud ovation, and newspaper coverage the following day was enthusiastic.

However successful, these demonstrations exposed remaining difficulties, particularly the need for a means of electrically amplifying weak signals, such as those coming from Governor Johnson's mansion, and using it to drive their loudspeaker. This was finally solved through the use of de Forest's audion tube, which had been used to spectacular effect by AT&T in 1915 in radio transmissions between Paris, Virginia, and Honolulu. Precisely how the rights to the audion were obtained is not clear, but this may have resulted from a private agreement with de Forest, who had sold his patent rights to AT&T but retained permission to sell equipment to "amateurs"—a term that de Forest construed quite liberally.

Having crossed this technological hurdle, the two inventors still faced other problems. The San Francisco demonstrations did not lead immediately to marketing opportunities, causing Jensen and Pridham to suspect that public speakers looked on the loudspeaker as an aspersion on their histrionic abilities. As the two inventors and their financial backers pondered an apparent lack of interest among prospective users of public-address systems, they decided that adapting the loudspeaker for use in an electronically amplified phonograph might have better market appeal. This, though, would require additional capital and a familiarity with the phonograph industry that neither Jensen and Pridham nor their backers possessed. Although efforts to interest the Columbia and Victor phonograph companies in their device proved fruitless, a chance encounter with

George Cook Sweet, a naval officer who visited their Napa laboratory and was impressed with the amplifying power of their equipment, led to a crucial contact with Frank M. Steers, president of the Sonora Phonograph Corporation in San Francisco. Late in 1916, Jensen and Pridham sold the Napa laboratory and moved to that city. On August 3, 1917, the CWDC was consolidated with Sonora and renamed the Magnavox Company. Although O'Connor remained a director in the new enterprise, Steers became its chief executive, while Jensen and Pridham were jointly designated chief engineer.

By this time the United States had entered World War I, delaying the phonograph gambit and leading Magnavox to confine itself to military applications. This opened up new prospects for the loudspeaker. In cooperation with the U.S. Navy, the company took part in an unsuccessful attempt to use the new device to facilitate communication between a ground station and plane flying overhead. Much more successful was a project in which, by stripping a microphone of its casing and mouthpiece, and exposing its diaphragm to the open air, Jensen and Pridham developed an antinoise transmitter that made the human voice audible over the roar of an airplane engine. Communication within an airplane during flight was permitted by using earphones encased in rubber inside the helmets of crew members. Widely used during the remainder of the war, the system saw service in the Curtiss NC-4 flying boats used in the U.S. Navy's pioneering transatlantic flight from Newfoundland to Lisbon in 1919. Jensen and Pridham also developed a public-address system for use on naval vessels, particularly amid the noise of engine rooms or during storms.

After the return of peace in 1918, the Magnavox Company was kept busy making watertight telephones for use on ships, and little was done to exploit the promise inherent in the loudspeaker. This suddenly changed, however, when President Woodrow Wilson came to California during his nationwide campaign to promote the American ratification of the Treaty of Versailles and membership in the League of Nations. Forbidden by his physicians to speak outdoors because of his failing health, it appeared that he would not be able to make a scheduled address in San Diego. Seeking to forestall the cancellation, a group of citizens approached the Magnavox Company about the possibility of installing a loudspeaker system in the municipal stadium so that Wilson could speak from a glass-enclosed booth. Magnavox agreed, and appropriate preparations were made.

Because Jensen was in the East on a business trip and could not attend the event, which took place on August 19, 1919, Pridham was put in charge of what turned out to be a briefly harrowing assignment. Playing recorded music to entertain the crowd of fifty thousand while Wilson's motorcade was taking part in a parade en route to the stadium, Pridham was horrified when the system went dead at the very moment the president's automobile came into view. As smoke began coming out of the amplifier, Pridham desperately removed the nearest tube from its socket and replaced it with a new one. Providentially, it worked; somehow he had gone directly to the source of the short circuit that had caused the difficulty. The following day, articles in such newspapers as the Chicago *Tribune* reported how Wilson, speaking in a normal tone of voice, was heard distinctly through two Magnavox speakers "with as much ease as if he were talking in a small banquet room."

Practically overnight, the loudspeaker now received rapid national as well as international attention and acceptance. Loudspeakers were used at both the Democratic and Republican national conventions in 1920, and both presidential candidates, James M. Cox and Warren G. Harding, used them extensively in their campaigns. A loudspeaker system was also used at Harding's inauguration in Washington, D.C., on March 4, 1921. Suddenly the market for loudspeakers appeared limitless, forcing Magnavox

Jensen in his laboratory in 1917. The noise-neutralizing microphone he is wearing was invented and patented by Jensen and Pridham.

to reassess its future. Company leaders feared that they could not meet the mounting demand with the capital at their disposal and also faced the inevitable consequences of a struggle with AT&T, which along with General Electric had a commanding position in vacuum tube technology. The California firm's West Coast location posed an additional problem, making it difficult to compete in key eastern markets. Magnavox therefore yielded most of the market for public-address systems to AT&T. Reverting to a strategy already formulated in 1917, from 1922 onward it devoted itself chiefly to the booming radio and phonograph industry, producing a succession of improved horn speakers, receivers, and vacuum tubes.

Now famous and with his financial position secure, Jensen went to Europe in 1922 to study developments in loudspeaker technology taking place there, develop new markets for Magnavox products in England, and visit his native land, where his mother still lived. Spending the summer in London, he staged the first demonstration of the loudspeaker ever to take place in that city in connection with a prizefight between George Carpenter and Ted "Kid" Lewis, which resulted in a scoop for his sponsor, the *Daily Mail*. He also met the father of radio, Guglielmo Marconi, for the first time, and chatted with him about Poulsen and the early days of wireless communication. Proceeding to Denmark, Jensen demonstrated his loudspeaker system to a throng gathered in City Hall Square in Copenhagen, creating a sensation when he amplified a long-distance telephone conversation with a correspondent in Stockholm so that the entire crowd could hear it.

Finally, Jensen went back to Falster, where his mother was taken aback by the unexpectedly resounding kiss with which he greeted her. "I had learned to show my affections in America," he later explained. "I hear you are famous," she responded. "Now I only hope it will last!" During his visit, he tramped around the island, reminiscing about old times. Gazing at the Strait of Groensund from a cliff overlooking the Baltic Sea, he was overcome by intense feelings of love for his native land. It made him realize that although he had "striven to throw off every vestige of a Dane in my effort to become a better citizen of my adopted country," he had become both "a better American and a better Dane than I had been before."

Returning to America, Jensen ultimately broke with other Magnavox executives over policy matters. Though Pridham remained with the company, Jensen left it in 1925. In 1927, he founded the Jensen Radio Manufacturing Company; originally headquartered in Oakland, California, it was soon moved to Chicago, where with the aid of engineer Hugh Knowles, Jensen worked intensively to eliminate distortion and improve fidelity in sound reproduction. In this new phase of his career, Jensen's marriage to his wife ended in divorce; in 1929, he married Malvena Opliger, by whom he had one child. In 1943, disputes with financial backers led once more to his resignation from a firm of his own creation, and he subsequently founded Jensen Industries to manufacture phonograph needles. He also served as chief consultant to the U.S. War Production Board's radio and radar division in World War II.

Jensen's old age brought him many honors, perhaps the most important being the conferring of knighthood by the king of Denmark in 1956. He also received awards from the American Institute of Radio Engineers and the Audio Engineering Society. A man of many interests, he studied philosophy, played the violin, enjoyed light opera, and took part avidly in various sports, particularly tennis. An inveterate smoker, he died of lung cancer in Western Springs, Illinois, on October 26, 1961. His friendship with Pridham, who died in 1963, had continued unabated throughout the years.

Jensen's career was aptly summarized by an obituary in the *Journal of the Audio Engineering Society*: "Of the many men who distinguished themselves in the early years of radio engineering, Peter Jensen was undoubtedly one of the most creative and productive. His inventions, experiments and sound systems were the heralds of today's high fidelity industry." In a broader perspective, Jensen's career provided a textbook case of how an important innovation may proceed out of a rich background of technological effort—in this case, the worldwide pattern of engineering activity that accompanied the emergence of radio in the late nineteenth and early twentieth centuries, and by going through characteristic stages of invention and development, yield a product that through commercial exploitation, can become a key part of everyday life. Jensen's lifetime spanned a period in which a long list of innovations including the telephone, the phonograph, motion pictures, radio, and television transformed human communications; the loudspeaker figured significantly in enhancing the power and impact of every one of them. In human terms, the life of this Danish American is a case study of the path followed by a legion of immigrant inventors, including such persons as Alexander Graham Bell, Michael Pupin, Nikola Tesla, Charles P. Steinmetz, and Vladimir Zworykin, who found the United States to be truly a land of opportunity and contributed to a process of technological transformation that continues to unfold in the late twentieth century.

Bibliograpical Note

Jensen's life can be traced in his autobiography, *En verdenskendt Dansker, Jensen, hottalerens opfinder: En selvbiografi* (Copenhagen: C. Erichsen, 1948), subsequently published in English as *The Great Voice* (Richardson, TX: Havilah Press, 1975). I have drawn heavily on both versions, and wish to thank Professor Ole Krogh, formerly of Auburn University, for his assistance in working with the Danish original. Supplementary information on Jensen's life is available in standard Danish biographical dictionaries, including *Dansk Biografisk Leksikon* (1937) and *Kraks blaa bog* (1949, 1961). For a brief synopsis of his career, see W. David Lewis, "Jensen, Peter Laurits," in *Dictionary of American Biography*, ed. John A. Garraty (New York: Charles Scribner's Sons, 1981), supplement 7 (1961–1965).

For authoritative analyses of the background of radio throughout Jensen's career, see Hugh G. J. Aitken, *Syntony and Spark: The Origins of Radio* (Princeton, NJ: Princeton University Press, 1976); Hugh G. J. Aitken, *The Continuous Wave: Technology and American Radio, 1900–1932* (Princeton, NJ: Princeton University Press, 1985). I greatly appreciate Aitken's help and advice as this chapter went through successive drafts; thanks are also due to Bernard Finn and Elliot Sivowitch of the National Museum of American History, the Scandinavian section of the Library of Congress, and Boyd Childress of the Ralph Brown Draughon Library of Auburn University for help at various stages of my research on Jensen's career. On the invention-development-innovation model of technological change, see Peter F. Drucker, "Technological Trends in the Twentieth Century," in *Technology in Western Civilization*, ed. Melvin Kranzberg and Carroll W. Pursell (New York: Oxford University Press, 1967); Devendra Sahal, *Patterns of Techno-logical Innovation* (Reading, MA: Addison-Wesley, 1981); John M. Stauden-maier, *Technology's Storytellers: Reweaving the Human Fabric* (Cambridge, MA: MIT Press, 1985), usefully supplemented by a series of articles by Thomas P. Hughes, Lynwood Bryant, Thomas M. Smith, Richard G. Hewlett, and Charles Susskind in a special section on "The Development Phase of Tech-nological Change" in *Technology and Culture* 17 (July 1976). On the shop culture model into which Jensen's training under Poulsen fits, see Monte

A. Calvert, *The Mechanical Engineer in America, 1830–1910: Professional Cultures in Conflict* (Baltimore: Johns Hopkins Press, 1967). On the development of the American broadcasting industry during the period in which Jensen lived, see particularly Erik Barnouw, *A History of Broadcasting in the United States*, 3 vols. (New York: Oxford University Press, 1966–1970); Susan J. Douglas, *Inventing American Broadcasting, 1899–1922* (Baltimore: Johns Hopkins University Press, 1987).

For additional perspectives on Jensen's career, see Glenn D. Kittler, "Forgotten Man of Sound," *Coronet*, May 1954; Jane Morgan, *Electronics in the West: The First Fifty Years* (Palo Alto, CA: National Press Books, 1967); Lillian C. White, *Pioneer and Patriot: George Cook Sweet, Commander, U.S.N.* (Delray Beach, FL: Southern Publishing Company, 1963); Robert Lozier, "Twenty Years of the Magnavox Story: 1911–1931," an undated reprint from the *Bulletin* of the Antique Wireless Association supplied to me by Aitken. An obituary of Jensen is in the *Journal of the Audio Engineering Society*, January 1961.

20   Charles A. Lindbergh: His Flight and the American Ideal*

John William Ward

On Friday, May 20, 1927, at 7:52 a.m., Charles A. Lindbergh took off in a silver-winged monoplane and flew from the United States to France. With this flight, Lindbergh became the first man to fly alone across the Atlantic Ocean. The log of flight 33 of the *Spirit of St. Louis* reads, "Roosevelt Field, Long Island, New York, to Le Bourget Aerodrome, Paris, France. 33 hrs. 30 min." Thus was the fact of Lindbergh's achievement easily noted. But the meaning of Lindbergh's flight lay hidden in the next sentence of the log: "(Fuselage fabric badly torn by souvenir hunters.)"

When Lindbergh landed at Le Bourget, he is supposed to have said, "Well, we've done it." A contemporary writer asked, "Did what?" Lindbergh "had no idea of what he had done. He thought he had simply flown from New York to Paris. What he had really done was something far greater. He had fired the imagination of mankind." From the moment of Lindbergh's flight, people recognized that something more was involved than the mere fact of the physical leap from New York to Paris. "Lindbergh," wrote John Erskine, "served as a metaphor." But what the metaphor stood for was not easy to say. The *New York Times* remarked then that "there has been no complete and satisfactory explanation of the enthusiasm and acclaim for Captain Lindbergh." Looking back on the celebration of Lindbergh, one can see now that the American people were trying to understand Lindbergh's flight, grasp its meaning, and through it, perhaps, grasp the meaning of their own experience. Was the flight the achievement of a heroic, solitary, unaided individual? Or did the flight represent the triumph of the machine, the success of an industrially organized society? These questions

* This essay, originally titled "The Meaning of Lindbergh's Flight," is reprinted by the courtesy of *The American Quarterly* from its Spring, 1958 issue.

were central to the meaning of Lindbergh's flight. They were also central to the lives of the people who made Lindbergh their hero.

The flight demanded attention in its own right, of course, quite apart from whatever significance it might have. Lindbergh's story had all the makings of great drama. Since 1919, there had been a standing prize of $25,000 to be awarded to the first aviator who could cross the Atlantic in either direction between the United States and France in a heavier-than-air craft. In spring 1927, there promised to be what the *New York Times* called "the most spectacular race ever held—3,600 miles over the open sea to Paris." Veteran pilots dominated the scene. On the European side were the French aces, Charles Eugène Jules Marie Nungesser and François Coli; on the American side, Commander Richard E. Byrd, in a big trimotored Fokker monoplane, led a group of contestants. Besides Byrd, who had already flown over the North Pole, there were Commander Noel Davis, flying a ship named in honor of the American Legion, which had put up $100,000 to finance his attempt; Clarence Chamberlin, who had already set a world's endurance record of more than fifty-one hours in the air in a Bellanca trimotored plane; and Captain René Fonck, the French war ace, who had come to America to fly a Sikorsky aircraft. The hero was unheard of and unknown. He was on the West Coast supervising the construction of a single-engine plane to cost only $10,000.

Then fate played its part. It seemed impossible that Lindbergh could get his plane built and east to New York in time to challenge his better-equipped and more famous rivals. Yet in quick succession, a series of disasters cleared his path. On April 16, Commander Byrd's *America* crashed on its test flight, crushing the leg of Floyd Bennett, who was one of the crew, and injuring Byrd's hand and wrist. On April 24, Chamberlin cracked up in his Bellanca, not seriously, but enough to delay his plans. Then on April 26, Commander Davis and his copilot lost their lives as the *American Legion* crashed on its final test flight. In ten days, accidents had stopped all of Lindbergh's American rivals. Nungesser and Coli, however, took off in their romantically named ship, the *White Bird*, from Le Bourget on May 8. The world waited, and Lindbergh, still on the West Coast, decided to try to fly the Pacific. But Nungesser and Coli were never seen again. As rumors filled the newspapers, as reports came in that the *White Bird* was seen over Newfoundland, Boston, and the Atlantic, it soon became apparent that Nungesser and Coli had failed, dropping to their death in some unknown grave. Disaster had touched every ship entered in the transatlantic race.

Now with the stage cleared, Lindbergh entered. He swooped across the continent in two great strides, landing only in Saint Louis. The first leg of his flight established a new distance record, but all eyes were on the Atlantic and the feat received little notice. Curiously, the first time Lindbergh appeared in the headlines of the New York papers was Friday, May 13. By this time Byrd and Chamberlin were ready once again, but the weather had closed in and kept all planes on the ground. Then after a week of fretful waiting, on the night of May 19 on the way into New York to see the Broadway musical *Rio Rita*, Lindbergh received a report that the weather was breaking over the ocean. He hurried back to Roosevelt Field to haul his plane out onto a wet, dripping runway. After mechanics painfully loaded the plane's gas by hand, the wind shifted, as fate played its last trick: a muddy runway and adverse wind. Whatever the elements, whatever the fates, the decisive act is the hero's, and Lindbergh made his choice. Providing a chorus to the action, the *Herald Tribune* reported that Lindbergh lifted the overloaded plane into the sky "by his indomitable will alone."

The parabola of the action was as clean as the arc of Lindbergh's flight. The drama should have ended with the landing of the *Spirit of St. Louis* at Le Bourget. That is where Lindbergh wanted it to end. In *"WE,"* written immediately after the flight, and *The Spirit of St. Louis*, written twenty-six years later, Lindbergh chose to end his accounts there. But the flight turned out to be only the first act in the part Lindbergh was to play.

Lindbergh was so innocent of his future that on his flight, he carried letters of introduction. The hysterical response, first of the French and then his own countrymen, had been no part of his careful plans. In *"WE,"* after Lindbergh's narrative of the flight, the publisher wrote, "When Lindbergh came to tell the story of his welcome at Paris, London, Brussels, Washington, New York, and St. Louis, he found himself up against a tougher problem than flying the Atlantic." So another writer completed the account in the third person. He suggested that "the reason Lindbergh's story is different is that when his plane came to a halt on Le Bourget field that black night in Paris, Lindbergh the man kept on going. The phenomenon of Lindbergh took its start with his flight across the ocean; but in its entirety it was almost as distinct from that flight as though he had never flown at all."

Acclaimed as the "lone eagle," Charles A. Lindbergh poses beside his monoplane, the *Spirit of St. Louis*.

A crowd gathers in front of the *Spirit of St. Louis* at the Le Bourget Airfield after Lindbergh's solo nonstop flight on May 20–21, 1927.
Source: UPI

Lindbergh's private life ended with his flight to Paris. The drama was no longer his; it belonged to the public. "The outburst of unanimous acclaim was at once personal and symbolic," said the *American Review of Reviews*. From the moment of success there were two Lindberghs: the private Lindbergh, and the public Lindbergh. The latter was the construction of the imagination of Lindbergh's time, fastened onto an unwilling person. The tragedy of Lindbergh's career is that he could never accept the role assigned him. He always believed he might keep his two lives separate. But from the moment he landed at Le Bourget, Lindbergh became, as the *New Republic* noted, "*ours.* ... He is no longer permitted to be himself. He is US personified. He is the United States." Ambassador Myron T. Herrick introduced Lindbergh to the French, saying, "This young man from out of the West brings you better than anything else the spirit of America," and wired to President Calvin Coolidge, "Had we searched all America we could not have found a better type than young Lindbergh to represent the spirit and high purpose of our people." This was Lindbergh's fate: to be a type. A writer in the *North American Review* felt that Lindbergh represented "the dominant American character," that he "images the best" about the United States. And an ecstatic female in the *American Magazine*, who began by saying that Lindbergh "is a sort of symbol. ... He is the dream that is in our hearts," concluded that the American public responded so wildly to Lindbergh because of "the thrill of possessing, in him, our dream of what *we* really and truly want to be." The act of possession was so complete that articles since have attempted to discover the "real" Lindbergh, that enigmatic and taciturn figure behind the public mask. Yet it is no less difficult to discern the features of the public Lindbergh, that symbolic figure who presented to the imagination of his time all the yearnings and buried desires of its dream for itself.

Lindbergh's flight came at the end of a decade marked by social and political corruption along with a sense of moral loss. The heady idealism of the First World War had been succeeded by a deep cynicism as to the war's real purpose. The naive belief that virtue could be legislated was violated by the vast discrepancy between the law and the social habits of Prohibition. A philosophy of relativism had become the uneasy rationale of a nation that had formerly believed in moral absolutes. The newspapers agreed that Lindbergh's chief worth was his spiritual and moral value. His story was held to be "in striking contrast with the sordid unhallowed themes that have for months steeped the imaginations and thinking of the people." Or as another had it, "There is good reason why people should hail Lindbergh and give him honor. He stands out in a grubby world as an inspiration."

Lindbergh gave the American people a glimpse of what they liked to think themselves to be at a time when they feared they had deserted their own vision of themselves. The grubbiness of the 1920s had a good deal to do with the shining quality of Lindbergh's success, especially when one remembers that Lindbergh's flight was not as unexampled as our national memory would have it. The Atlantic was not unconquered when Lindbergh flew. A British dirigible had twice crossed the Atlantic before 1919, and on May 8 of that year, three naval seaplanes left Rockaway, New York, and one, the NC-4 manned by a crew of five, got through to Plymouth, England. A month later, Captain John Alcock, an Englishman, with Arthur W. Browne, an American, flew the first heavier-than-air land plane across the Atlantic nonstop, from Newfoundland to Ireland, to win twice the money Lindbergh did: a prize of $50,000 offered by the London *Daily Mail*. Alcock and Browne's misfortune was to land in a soft and somnolent Irish peat bog instead of before the cheering thousands of London or Paris. Or perhaps they should have flown in 1927.

The wild medley of public acclaim and Homeric strivings of editors make one realize that the response to Lindbergh involved a mass ritual in which America celebrated itself more than it celebrated Lindbergh. Lindbergh's flight was the occasion of a public act of regeneration in which the nation momentarily rededicated itself to something, the loss of which was keenly felt. It was said again and again that "Lindy" taught America "to lift its eyes up to Heaven." Heywood Broun, in his column in the *New York World*, wrote that this "tall young man raised up and let us see the potentialities of the human spirit." Broun felt that the flight proved that though "we are small and fragile," it "isn't true that there is no health in us." Lindbergh's flight provided the moment, but the meaning of the flight is to be found in the deep and pervasive need for renewal that the flight brought to the surface of public feeling. When Lindbergh appeared in the nation's capital, the *Washington Post* observed, "He was given that frenzied acclaim which comes from the depths of the people." In New York, where four hundred thousand people saw him, a reporter wrote that the dense and vociferous crowds were swept, as Lindbergh passed, "with an emotion tense and inflammable." The

New York welcomes Lindbergh with a triumphal ticker tape parade on June 14, 1927.
Source: UPI

Literary Digest suggested that the answer to the hero worship of Lindbergh would "throw an interesting light on the psychology of our times and of the American people."

The *Nation* noted about Lindbergh that

there was something lyric as well as heroic about the apparition of this young Lochinvar who suddenly came out of the West and who flew all unarmed and all alone. It is the kind of stuff which the ancient Greeks would have worked into a myth and the medieval Scots into a border ballad. ... But what we have in the case of Lindbergh is an actual, an heroic and an exhaustively exposed experience which exists by suggestion in the form of poetry.

The *Nation* quickly qualified its statement by observing that reporters were as far as possible from being poets, and concluded that the discrepancy between the fact and celebration of it was not poetry perhaps but nonetheless "magic on a vast scale." Yet the *Nation* might have clung to its insight that the public meaning of Lindbergh's flight was somehow poetic. The vast publicity about Lindbergh corresponds in one vital particular with the poetic vision. Poetry, said William Butler Yeats, contains opposites; so did Lindbergh. Lindbergh did not mean one thing; he meant many things. The image of itself that America contemplated in the public person of Lindbergh was full of conflict; it was, in a word, dramatic.

To heighten the drama, Lindbergh did it alone. He was the "lone eagle," and a full exploration of that fact takes one deep into the emotional meaning of his success. Not only the *Nation* found Sir Walter Scott's lines on Lochinvar appropriate: "he rode all unarmed and he rode all alone." Newspapers and magazines were deluged with amateur poems that vindicated one rhymester's wry comment, "Go conquer the perils / That lurk in the skies— / And you'll get bum poems / Right up to your eyes." The *New York Times*, which alone received more than two hundred poems, observed in trying to summarize the poetic deluge that "the fact that he flew alone made the strongest impression." Another favorite tribute was Rudyard Kipling's "The Winners," with its refrain, "He travels the fastest who travels alone." The others who had conquered the Atlantic, and those like Byrd and Chamberlin who were trying at the same time, were not traveling alone, and they hardly rode unarmed. Other than Lindbergh, all the contestants in the transatlantic race had unlimited backing and access to the best planes, and all were working in teams, carrying at least one copilot to share the long burden of flying the plane. So a writer in the New York *Sun*, in a poem called "The Flying Fool," a nickname that Lindbergh despised, celebrated Lindbergh's flight: "No kingly plane for him; / No endless data, comrades, moneyed chums; / No boards, no councils, no directors grim— / He plans ALONE … and takes luck as it comes."

On second thought, it must seem strange that the long-distance flight of an airplane, the achievement of a highly advanced and organized technology, should be the occasion for hymns of praise to the solitary unaided man. Yet the National Geographic Society, when it presented a medal to Lindbergh, wrote on the presentation scroll, "Courage, when it goes alone, has ever caught men's imaginations," and compared Lindbergh to Robinson Crusoe and the trail makers in our own West. But Lindbergh and Crusoe, the one in his helmet and fur-lined flying coat, and the other in his wild goatskins, do not easily coexist. Even if Crusoe did have a tidy capital investment in the form of a well-stocked shipwreck, he still did not have a $10,000 machine under him.

Lindbergh, in nearly every remark about his flight and his own writings about it, resisted the tendency to exploit the flight as the achievement of an individual. He never said "I"; he always said "we." The plane was not to go unrecognized. Nevertheless, there persisted a tendency to seize on the flight as a way of celebrating the self-sufficient individual, so that among many others, an Ohio newspaper could describe Lindbergh as this "self-contained, self-reliant, courageous young man [who] ranks among the great pioneers of history." The strategy here was a common one: to make Lindbergh a "pioneer," and thus link him with a long and vital tradition of individualism in the American experience. Colonel Theodore Roosevelt, himself the son of a famous exponent of self-reliance, said to reporters at his home in Oyster Bay that "Captain Lindbergh personifies the daring of youth. Daniel Boone, David Crocket [sic], and men of that type played a lone hand and made America. Lindbergh is their lineal descendant." In *Outlook* magazine, immediately below an enthusiastic endorsement of Lindbergh's own remarks on the importance of his machine and his scientific instruments, there was the statement, "Charles Lindbergh is the heir of all that we like to think is best in America. He is of the stuff out of which have been made the pioneers that opened up the wilderness, first on the Atlantic coast, and then in our great West. His are the qualities which we, as a people, must nourish." It is in this mood that one suspects it was crucial that Lindbergh came out of the West and rode alone.

Another common metaphor in the attempt to place Lindbergh's exploit was to say that he had opened a new "frontier." To speak of the air as a frontier was to invoke an interpretation of the meaning of American history that had sources deep in the American experience, but the frontier of the airplane is hardly that of the trail makers of the old West. Rather than an escape into the self-sufficient simplicity of the American past, the machine that made Lindbergh's flight possible represented an advance into a complex industrial present. The difficulty lay in using an instance of modern life to celebrate the virtues of the past, to use an extreme development of an urban industrial society to insist on the significance of the frontier in American life.

A little more than a month after Lindbergh's flight, Joseph K. Hart in *Survey* magazine reached back to Walt Whitman's poem for the title of an article on Lindbergh: "O Pioneer." A school had made Lindbergh an honorary alumnus, but Hart protested there was little available evidence "that he was educated in *schools*." "We must look elsewhere for our explanation," Hart wrote, and he looked to the experience of Lindbergh's youth when "everything that he ever did ... he did by himself. He lived more to himself than most boys." And of course, Lindbergh lived to himself in the only place conceivably possible on a Minnesota farm: in the world of nature. "There he developed in the companionship of woods and fields, animals and machines, his audaciously natural and simple personality," The word "machines" jars as it intrudes into Hart's idyllic pastoral landscape and betrays Hart's difficulty in relating the setting of nature on which he wishes to insist with the fact that its product spent his whole life tinkering with machines, from motorcycles to airplanes. But except for that one word, Hart proceeds in uncritical nostalgia to show that "a lone trip across the Atlantic was not impossible for a boy who had grown up in the solitude of the woods and waters." If Lindbergh was "clear-headed, naif, untrained in the ways of cities," it was because he had "that 'natural simplicity' which Fenimore Cooper used to attribute to the pioneer hero of his Leatherstocking Tales." Hart rejected the notion that any student "bent to all the conformities" of formal training could have done what Lindbergh did. "Must we not admit," he asked, "that this pioneering urge remained to this audacious youth because he had never submitted completely to the repressions of the world and its jealous institutions?"

Only those who insist on reason will find it strange that Hart should use the industrial achievement of the airplane to reject the urban, institutionalized world of industrialism. Hart was dealing with something other than reason; he was dealing with the emotion evoked by Lindbergh's solitude. He recognized that people wished to call Lindbergh a "genius" because that "would release him from the ordinary rules of existence." That way, "we could rejoice with him in his triumph, and then go back to the contracted routines of our institutional ways [because] 99 percent of us must be content to be shaped and moulded by the routine ways and forms of the world to the routine tasks of life." It is in the word "must" that the pathos of this interpretation of the phenomenon of Lindbergh lies. The world had changed from the open society of the pioneer to the close-knit, interdependent world of a modern machine-oriented civilization. The institutions of a highly corporate industrial society existed as a constant reproach to a people who liked to believe that the meaning of its experience was embodied in the formless, independent life of the frontier. Like Thomas Jefferson, who identified American virtue with nature and saw the city as a "great sore" on the public body, Hart concluded,

Certainly, in the response that the world—especially the world of great cities—has made to the performance of this midwestern boy, we can read of the homesickness of the human soul, immured in city canyons and routine tasks, for the freer world of youth, for the open spaces of the pioneer, for the joy of battling with nature and clean storms once more on the frontiers of the earth.

The social actuality that made the adulation of Lindbergh possible had its own irony for the notion that America's strength was in its simple, uncomplicated beginnings. For the public response to Lindbergh to have reached the proportions it did, the world had by necessity to be the intricately developed world of modern mass communications. But more than irony was involved. Ultimately, the emotion attached to Lindbergh's flight involved no less than a whole theory about American history. By singling out the fact that Lindbergh rode alone and naming him a pioneer of the frontier, the public projected its sense that the source of America's strength lay somewhere in the past and Lindbergh somehow meant that America must look backward in time to rediscover some lost virtue. The mood was nostalgic, and American history was read as a decline—a decline measured in terms of America's advance into an urban, institutionalized way of life that made solitary achievement increasingly beyond the reach of 99 percent of the people. Because Lindbergh's ancestors were Norse, it was easy to call him a "Viking" and extend the emotion far into the past when all frontiers were open. He became the "Columbus" of another new world to conquer as well as the "Lochinvar" who rode all alone. There was always the brute, irreducible fact, however, that Lindbergh's exploit was a victory of the machine over the barriers of nature. If the only response to Lindbergh had been a retreat to the past, we would be involved with a mass cultural neurosis, the inability of America to accept reality, the reality of the world in which it lived. But there was another aspect, one in which the public celebrated the machine and the highly organized society of which it was a product. The response to Lindbergh reveals that the American people were deeply torn between conflicting interpretations of their own experience. By calling Lindbergh a pioneer, the people could read into American history the necessity of turning back to the frontier past. Yet the people could also read American history in terms of progress into the industrial future. They could do this by emphasizing the machine that was involved in Lindbergh's flight.

Lindbergh came back from Europe in an American man-of-war, the cruiser *Memphis*. It seems he had contemplated flying on, around the whole world perhaps, but less adventurous heads prevailed and dictated a surer mode of travel for so valuable a piece of public property. The *New Republic* protested against bringing America's hero of romance home on a warship. If he had returned on a great liner, that would have been one thing. "One's first trip on an oceanliner is a great adventure—the novelty of it, the many people of all kinds and conditions, floating for a week in a tiny compact world of their own." But to return on the *Memphis*, "to be put on a gray battleship with a collection of people all of the same stripe, in a kind of ship that has as much relation to the life of the seas as a Ford factory has! We might as well have put him in a pneumatic tube and shot him across the Atlantic." The interesting thing about the *New Republic's* protest against the unromantic, regimented life of a battleship is that the image it found appropriate was the Ford assembly line. It was this reaction against the discipline of a mechanized society that probably led to the nostalgic image of Lindbergh as a remnant of a past when romance was possible for the individual, when life held novelty and society was variegated rather than uniform. What the Ford assembly line represents, though—a society committed to the path of full mechanization—was what lay behind Lindbergh's romantic success. A long piece in the Sunday *New York Times*, "Lindbergh Symbolizes the Genius of America," reminded its readers of the too-obvious fact that "without an airplane he could not have flown at all." Lindbergh "is, indeed, the Icarus of the twentieth century; not himself an inventor of his own wings, but a son of that omnipotent Daedalus whose ingenuity has created the modern world." The point was that modern America was the creation of modern industry. Lindbergh "reveres his 'ship' as a noble expression of mechanical wisdom. ... Yet in this reverence ... Lindbergh is not an exception. What he means by the Spirit of St. Louis is really the spirit of America. The mechanical genius, which is discerned in Henry Ford as well as in Charles A. Lindbergh, is in the very atmosphere of [the] country." In contrast to a sentiment that feared the enforced discipline of the machine, there existed an attitude of reverence for its power.

Lindbergh led the way in the celebration of the machine, not only implicitly by including his plane when he said "we," but also by direct statement. In Paris he told newspapermen, "You fellows have not said enough about that wonderful motor." Rarely have two more taciturn figures confronted one another than when Lindbergh returned to Washington, D.C., and Coolidge pinned the Distinguished Flying Cross on him, but in his brief remarks, Coolidge found room to express his particular delight that Lindbergh should have given equal credit to the airplane. "For we are proud," said the president, "that in every particular this silent partner represented American genius and industry. I am told that more than 100 separate companies furnished materials, parts or service in its construction."

The flight was not the heroic lone success of a single daring individual but rather the climax of the cooperative effort of an elaborately interlocked technology. The day after Coolidge's speech, Lindbergh said at another ceremony in Washington that the honor should "not go to the pilot alone but to American science and genius which had given years of study to the advancement of aeronautics." "Some things," he said, "should be taken into due consideration in connection with our flight that have not heretofore been given due weight. That is just what made this flight possible. It was not the act of a single pilot. It was the culmination of twenty years of aeronautical research and the assembling together of all that was practicable and best in American aviation." The flight, concluded Lindbergh, "represented American industry."

The worship of the machine that was embodied in the public's response to Lindbergh exalted those very aspects that were denigrated in the celebration of the flight as the work of a heroic individual. Organization and careful method were what lay behind the flight, not individual self-sufficiency and daring romance. One magazine hailed the flight as a "triumph of mechanical engineering. ... It is not to be forgotten that this era is the work not so much of brave aviators as of engineers, who have through patient and protracted effort been steadily improving the construction of airplanes." The lesson to be learned from Lindbergh's flight, thought a writer in the *Independent*, "is that the splendid human and material aspects of America need to be organized for the ordinary, matter of fact service of society." The machine meant organization, the careful rationalization of activity of a Ford assembly line; it meant planning, and if it meant the loss of spontaneous individual action, it meant the material betterment of society. Lindbergh meant not a retreat to the free life of the frontier past but instead an emergence into the time when "the machine began to take first place in the public mind—the machine and the organization that made its operation possible on a large scale." A poet on this side of the matter wrote, "All day I felt the pull / Of the steel miracle." The machine was not a devilish engine that would enthrall mankind; it was the instrument that would lead to a new paradise. Yet the direction of history implicit in the machine was toward the future, not the past; the meaning of history was progress, not decline, and America should not lose faith in the future betterment of society. An address by a Harvard University professor, picked up by the *Magazine of Business*, made all this explicit. "We commonly take Social Progress for granted," said Edwin F. Gay, "but the doctrine of Social Progress is one of the great revolutionary ideas which have powerfully affected our modern world." The idea, however, "may be in danger of becoming a commonplace or a butt of criticism." The speaker recognized why this might be. America was "worn and disillusioned after the Great War." Logically, contentment should have gone with so optimistic a creed, yet the American people were losing faith. So Lindbergh filled an emotional need even where a need should have been lacking. "He has come like a shining vision to revive the hope of mankind." The high ideals of faith in progress "had almost come to seem like hollow words to us—but now here he is, emblematic of heroes yet to inhabit this world. Our belief in Social Progress is justified symbolically in him."

It is a long flight from New York to Paris; it is a still longer one from the fact of Lindbergh's achievement to the burden imposed on it by the imagination of his time. But it is in that further flight that one can find the full meaning of Lindbergh. His role was finally a double one. His flight provided an opportunity for the people to project their own emotions into his act, and their emotions involved ultimately two attitudes toward the meaning of their own experience. One view had it that America represented a brief escape from the course of history, an emergence into a new and open world with the self-sufficient individual at its center. The other said that America represented a stage in historical evolution and its fulfillment lay in the development of society. For one, the meaning of America lay in the past; for the other, it was in the future. For one, the American ideal was an escape from institutions, from the forms of society and the limitations put on the free individual; for the other, the American ideal was the elaboration of the complex institutions that made modern society possible, an acceptance of the discipline of the machine, and the achievement of the individual within a context of which he was only a part. The two views were contradictory, but both were possible, and both were present in the public's reaction to Lindbergh's flight.

The Sunday newspapers announced that Lindbergh had reached Paris, and in the very issue whose front pages were covered with Lindbergh's story, the magazine section of the *New York Times* featured an article by the British philosopher Bertrand Russell. The magazine had, of course, been made up too far in advance to take advantage of the news about Lindbergh. Yet in a prophetic way, Russell's article was about Lindbergh. Russell hailed the rise to power of the United States because he felt that in the "new life that is America's" in the twentieth century, "the new outlook appropriate to machinery [would] become more completely dominant than in the old world." Russell sensed that some might be unwilling to accept the machine, but "whether we like this new outlook or not," he wrote, "is of little importance." Why one might not was obvious. A society built on the machine, said Russell, meant "the diminution in the value and independence of the individual. Great enterprises tend more and more to be collective, and in an industrialized world the interference of the community with the individual must be more intense." Russell realized that while the cooperative effort involved in machine technology makes man collectively lordlier, it makes the individual more submissive. "I do not see how it is to be avoided," he concluded.

People are not philosophers. They did not see how the conflict between a machine society and the free individual was to be avoided either. But neither were they ready to accept the philosopher's statement of the problem. In Lindbergh, the people celebrated both the self-sufficient individual and the machine. Americans still celebrate both. We cherish the individualism of the American creed at the same time that we worship the machine that increasingly enforces collectivized behavior. Whether we can have both—the freedom of the individual and power of an organized society—is a question that still haunts our minds. To resolve the conflict that is present in America's celebration of Lindbergh in 1927 is still the task of America.

21 Buster Keaton and Charlie Chaplin: The Silent Film's Response to Technology

George Basalla

Any discussion of movies and modern technology is certain to focus on Charlie Chaplin and the problems he encountered as a factory worker in the film *Modern Times* (1936). American cinematic reaction to technology, however, began prior to Chaplin's portrayal of a beleaguered factory hand, and it can be fully understood only if we consider the mechanical nature of film itself and the motion picture comic tradition that preceded *Modern Times*.

Modern Times, as well as any other film, comic or tragic, silent or sound, live or animated, must be considered in relation to one of the more significant developments of nineteenth-century culture: the mechanization of the arts. Beginning with the invention of photography in the 1840s, technology posed a threat to traditional artistic practices and forms, while at the same time it raised the possibility of novel techniques and wholly new art forms.

Arguments about the artistic status of photography and role of the photograph as a source of inspiration for the painter or sculptor had not subsided when the phonograph appeared in the 1870s. Although the phonograph did not immediately raise questions about the creation of a new art form, as did the camera, it could be cited with photography as proof that artistic works, no matter how subtle or sublime, were amenable to *mechanical reproduction*. The advent of the camera and phonograph proved that art could be multiplied by a machine and prepared for wide distribution to the masses.

The culmination of the movement toward the mechanization of the arts was reached with the invention of motion pictures in the final decades of the nineteenth century. In cinema, mechanical civilization found its natural mode of artistic expression. It was an art form that was in essence mechanical, industrial, and commercial—a fact that is reflected in the characterization of early films as "canned" drama or machine-made theater. Among other things, the motion picture is a modern industrial product that results from the cooperative efforts of a group of artists, entrepreneurs, and technicians. To extend the industrial metaphor, it has been said that these makers of movies assemble their product in a manner analogous to that used on an automobile assembly line. Prior to the development of the cinema, no art form had ever been so dependent on the mechanical and commercial world for its full realization.

Who were the first producers and consumers of the cinematic illusions created by the Hollywood film industry? The early movie moguls were, for the most part, self-made men, immigrant entrepreneurs who founded the film industry and then continued to dominate it throughout most of the twentieth century. Initially, they provided entertainment for an industrial working-class audience that could afford five or ten cents for a movie ticket, but not the dollar or two needed for admittance to a performance at a legitimate theater or opera.

Technology, so intimately associated with the creation of moving pictures and their early audience, also accounted for an important part of the actual stories and images of the film. Arnold Hauser in his *Social History of Art* claims that because cinema was founded in technology, it is at its best in describing "movement, speed, and pace." Silent filmmakers searched their rural and urban landscape for things that moved with speed, and they found them in abundance: steam locomotives, motorcycles, steamships, airplanes, trolley cars, and automobiles. Long before there was developed an explicit aesthetic of the machine, filmmakers made use of the engines of industrial civilization. They elevated machines to the center of cinematic interest and, in some notable cases, managed to make them the stars of the moving picture show.

Virtually all the early slapstick comedies, especially those of Mack Sennett, incorporated hectic chase scenes involving an automobile, locomotive, trolley, and so on. Some of the early filmmakers, entranced with the depiction of speed on the screen, offered films that were merely melodramatic, comic, or mechanical embellishments of a prolonged chase. Whether their goal was comedy or tragedy, the men who made silent films concentrated on moving objects so essential to their art form, and thus developed a cinema of speed and exhilaration that has remained a crucial part of modern cinematic art.

*   *   *

The foregoing discussion of the early technological background and associations of film art, audience, and content forms a proper introduction to a close examination of two dominant film personalities who dealt with the machine and industrial society in their comedies: Buster Keaton and Chaplin.

Throughout his life, Keaton was an inventor, designer, and builder of machines. Some of the devices were props for his films, others were constructed as practical jokes (an outhouse with collapsing walls), and still others were absurd or fanciful contraptions in the style of Rube Goldberg. Keaton's fantastic machines reveal his understanding of the mechanical—the gadgets he built worked—appreciation of mechanical motion for its own sake, and ability to use the machine as a source of gentle and muted satire of a society that went to great technological lengths in order to accomplish the simple or unnecessary.

Between 1924 and 1926, Keaton released two films that dealt directly with some major examples of technology. In *The Navigator* (1924), he

Buster Keaton sits undaunted as his automobile disintegrates around him.
Source: Film Stills Archives / Museum of Modern Art, New York

turned a large passenger liner into a nemesis that threatened the comic figure he created on the screen. A pampered, incompetent, young millionaire and his reluctant ladylove are cast adrift, alone, on a five-hundred-foot steam vessel called the *Navigator*. Without crew, fuel, or electricity, the ocean liner becomes a monster, a "mechanical whale," with Jonah-Keaton and friend trapped inside. Food is available in abundance, but packaged for two hundred, not two; modern cookware abounds, but its capacities are measured in gallons, not cups. The darkened ship at night is as ominous a setting as a haunted house. But resourceful Keaton, meeting the challenge of the machine, creates a hospitable environment aboard, especially in the kitchen, where a set of his ingenious inventions enabled the couple to get the food they needed.

Keaton's biographer, Rudi Blesh, has said that *The Navigator* is not the story of romance impeded by the machine but rather "the tale of a precarious wooing that hinges on [Keaton's] duel with a mechanical enemy." There is no doubt of the machine's hostility in *The Navigator*, but neither is there doubt that Keaton will deal successfully with technology's challenge to the individual. The little guy always manages to outwit the big machine.

The General (1926) suggested a different emphasis in Keaton's continuing search for a solution to the personal problems engendered by the machines people meet in their lives. Set in the Civil War period, the film was based on the true story of a daring but unsuccessful attempt by Northern raiders in 1862 to infiltrate Southern territory, and disrupt rail service between Georgia and Tennessee. The Northern plan included the capture of a locomotive, the *General*, which the raiders would use as they moved northward burning bridges and cutting communication lines.

Since the *General* was a Southern locomotive, Keaton saw the necessity of telling the story from the Confederate viewpoint in order to elicit sympathy for engine and engineer. Keaton was best able to exploit the full cinematic and comic possibilities of the story not only because he expressed Confederate sympathies but also because he made his hero a civilian closely associated with the film's mechanical costar. This move called for a more harmonious relationship between man and machine. The *General* is no *Navigator* beyond the control of those aboard; it is a love object—normally obedient to man's will—that is lost and recovered.

The *General* and Keaton's girlfriend, Annebelle Lee, are both loved by him, yet their individual and collective actions can be frustrating to our hero. Neither is malicious, but neither is entirely cooperative. The girl is a bit scatterbrained, and the engine, often when under the girl's control, works contrary to Keaton's best interests and intentions. And when girl and engine cooperate, other mechanical malfunctions arise. Keaton is able to handle these malfunctions, but only in the nick of time.

It is not the machine's hostility toward people at issue here, it is that mechanical and human needs as well as desires do not mesh smoothly, they are slightly out of phase and require a Keaton, with wit and ingenuity, rather than superior technical knowledge, to set them right. All this might be summarized in a moral derived from Keaton's films: if people are to prevail over technology, they must be wary, even of the machines they love.

*   *   *

Keaton, who had no artistic successes with sound films, had shown a preference in his best works for historical settings—the Civil War—and machines of the nineteenth century—steam engines. His primary concern was not the social, economic, and technological problems of twentieth-century America. Chaplin, who gained fame in both silent and sound movies, made what is probably the most famous film commenting on twentieth-century technology: *Modern Times*.

Chaplin was motivated to make a film on the role of the machine in modern society by a young reporter who, while interviewing him, happened to describe working conditions in industrial Detroit. He was told that healthy, young farm boys were lured to the city to work on the assembly line producing automobiles. Within four or five years, these same young men were "nervous wrecks," their health destroyed by the pace of work set in the factory. The reporter's observations reinforced Chaplin's own "impulse to say something about the way life is being standardized and channelized, and men turned into machines."

Chaplin did not share Keaton's intimate knowledge of the machine, but he did aspire to, and move in, intellectual and social circles where technology might be the subject of abstract discussion. Silent films had made Chaplin an international figure with access to the greatest minds of the day. The combination of his own poverty-stricken childhood and the influence of the more socially concerned intellectuals he met, men like Max Eastman and H. G. Wells, turned multimillionaire Chaplin into a parlor socialist who enjoyed discussing social, economic, and political matters with radical overtones.

Apart from any expression it might find in *Modern Times*, what were Chaplin's attitude toward labor and the machine? The answer must be a confusing and unsatisfactory one, for it is derived from Chaplin's occasional excursions into the realms of economics and political philosophy. The standard Chaplin solution for all economic problems was to "reduce the hours of labor, print more money, and control prices." During the American Depression, he lectured Albert Einstein along these lines and was pleased when the physicist replied, "You are not a comedian. You're an economist." The Einstein conversation also found Chaplin claiming that technological unemployment—new machines displacing men from their jobs—was at the heart of America's economic problems in the 1930s. Yet when Chaplin had an interview with Mahatma Gandhi, he made clear his sympathy with the drive for India's independence but expressed confusion at the Indian leader's "abhorrence of machinery." Gandhi patiently explained to the comedian that India could only gain political independence if she first reduced her technological dependence on Great Britain. In all this there is nothing that is profound or novel. Chaplin was merely passing on some of the liberal clichés of his day, and doing it in a naive and superficial manner.

Modern Times, as described in its foreword by Chaplin, "is the story of industry, of individual enterprise—humanity crusading in the pursuit of happiness." The opening scene discloses a mass of sheep, jammed into pens, presumably on their way to the abattoir. Abruptly this scene cuts to a similar one of a crowd of factory workers jostling one another on their way to work in an industrial plant. The sharp note of social criticism implied by this sheep/worker montage is not sustained throughout the film. It is replaced by a more gentle tone of satire, by an obvious desire to please the audience and not to lecture to it.

Scenes of factory interiors account for less than one-third of the footage of *Modern Times*, but they contain some of the most pointed social commentary as well as the funniest comic situations. Everyone who has seen the film recalls Chaplin vainly trying to adapt to the pace of the fast-moving conveyor belt, literally going mad in the process, and running

Caught up in the giant cogs of a machine, Charlie Chaplin personifies the dehumanizing effects of mass production in *Modern Times* (1936).
Source: Film Stills Archives / Museum of Modern Art, New York

about the plant tightening real and imaginary nuts with his wrench as well as spraying oil on factory personnel.

Another popular scene featured a newly invented automatic feeding machine brought to the assembly line so the worker need not interrupt his labors while he ate. The feeding machine malfunctioned, hurling soup, meat, corn on the cob, and a napkin at Chaplin, who was strapped into the contraption and could not escape. In all of cinema, there is no finer scene illustrating man's utter helplessness before the machine that supposedly exists only to serve his most basic needs.

The bulk of *Modern Times* was set beyond factory walls in the institutions and business establishments of an industrial town. In this way, we learn not only about the assembly line but also about the social and home life of the workingman. The comedy continues in the new setting, but the scene is a grim, depressing one. Chaplin leaves the factory to recuperate in a hospital; from there he goes to prison. A strike by the factory workers brings with it violence, death, and unemployment—all of which impinges on Chaplin's life. Mechanical transportation for Chaplin takes the form of ambulances and police prison vans, not the automobile so dear to the American worker's dream. The one idyllic interlude in the film finds Chaplin with a young girl he befriends and protects. They play at housekeeping in an

abandoned waterfront shack, mocking middle-class respectability and pretensions. The intervention of the juvenile authorities seeking the girl ends
their happiness, and at the film's conclusion, Chaplin and friend tramp
down a lonely road into the sunset. They have been rebuffed and harassed
by every facet of industrial society; their only hope lies in escaping from it.
They leave behind them maddening work on the assembly line, unemployment, strikes, poverty, and social injustice. Chaplin had no ready solutions
to these problems.

Film critics of the day tended to respond to *Modern Times* according to
their preconceived notions of Chaplin and socialism. A writer for the Com-
munist *New Masses* praised it as the first American film "daring to challenge the superiority of an industrial civilization based upon the creed of
men who sit at flat topped desks and push buttons demanding more speed
from tortured employees." Others, more conservative, were relieved to find
that it was "neither fish, flesh, nor good red propaganda," that its social
message was innocuous and its comedy high, if at times uneven. Over the
years, critical opinion has tended to stress the derivative nature of *Modern
Times*—Chaplin lifted many comic bits from his earlier films—and its epi-
sodic character.

Clearly *Modern Times* has its faults, both of an artistic and ideological
sort, but it remains the single-best film ever made for a mass audience
treating technology within a social context. It did not offer a radical social
message for those wishing to overthrow capitalism; it did, however, accurately reflect the sentiments of many who felt they were helpless victims
of an overmechanized world. And finally, it did conceal a truly disturbing
message beneath its wonderful comic scenes: human beings cannot master
machines. The neat tricks that Keaton used to keep his machines in line
simply have no place in Chaplin's *Modern Times*. The only solution is to
turn one's back on the whole affair and walk down the road with a friend.
But over the next hill, we know that Chaplin is likely to find waiting for
him another factory, another industrial town, another set of problems generated by the social impact of technology.

Chaplin on the screen might decide to turn away from the machine in
desperation and seek happiness elsewhere. Yet Chaplin as a film director
had to determine exactly how he would handle the technology of sound
in *Modern Times*. The talkie was almost a decade old, and it had eclipsed

the silent film. In a New York Times article of 1931, Chaplin argued that talking pictures could never serve as adequate substitutes for silent films based on the universal language of pantomime. He ignored lip-synchronized sound when he made *City Lights* (1931) and went on to issue *Modern Times* five years later as the last full-length silent feature film made in Hollywood.

There is a nondialogue, musical soundtrack for *Modern Times*, but the film was shot for the most part at silent speed. At three specific points in the film, recognizable human speech was permitted to intrude into the otherwise-pantomimed action: first, when the factory's top executive used *closed-circuit television* to talk to his employees in the plant; second, when the inventor of the feeding machine played a *phonograph record* extolling the virtues of his device; and third, when a *radio* broadcasts a news bulletin in the prison scene. By limiting the sound of the human voice to three technological devices capable of transmitting and reproducing it, he underlined the technological character of the talkies that had supplanted the silent movies. The talking motion picture, he appeared to be saying, is but another technical gadget; the true value of a film's drama lies in the pantomimic art of the actors and actresses.

Even Chaplin, the master of pantomime, was forced to bow to what most people then saw as inevitable technological progress in the cinema. Near the conclusion of *Modern Times*, Chaplin sang a song that had deliberately contrived gibberish lyrics and a great deal of pantomime. Nevertheless, filmgoers heard Chaplin's voice for the first time over the loudspeakers in a movie theater. After that there could be no going back, no walking off into the sunset. If Chaplin were to produce another film, it must be a full-fledged talking picture. *The Great Dictator*, appearing four years later, was such a picture.

* * *

Technology was not a subject matter haphazardly introduced into the silent film by directors who were willing to include in their screen productions anything they happened on. The machine, in mechanizing the arts, gave birth to the cinema, which was especially suited to the needs and desires of industrial society; the machine deeply influenced film content and acting style; the machine brought sound to the silent movie and altered the future direction of cinematic development. Similarly, it was not by chance that the two greatest silent film comedians, Keaton and Chaplin, used cinema to explore the nature of technology, man, and society. They were using a mechanical art form in an attempt to understand the changing machine age in which they lived. What is ironic in all this is that a medium of artistic communication closely allied to the machine did not produce film master-pieces praising technology but instead ones that were suspicious and criti-cal of the machine along with the civilization it spawned.

22 Morris L. Cooke and Energy for America

Jean Christie

On September 4, 1882, on Pearl Street in Lower Manhattan, the first central electric power station sent current out over its wires. Its customers were owners of some four hundred of the carbon filament light bulbs recently invented by Thomas A. Edison, but Edison, founder and director of the company, insisted that electricity would soon find a market not only for lighting but many other uses as well. Not long afterward, advances in the transmission of alternating current enlarged the territory of central stations, and within thirty or forty years electric light superseded gas, electric trolley cars transported passengers within and between the growing cities, and electric motors replaced moving belts in the factories of rapidly industrializing America. Although coal-burning plants generated most of the current, electric companies, attracted by the potential of hydroelectric power, also sought to obtain promising sites on the nation's rivers. Engineers constructed larger plants and longer transmission lines, and bankers, entrepreneurs, and speculators organized complex structures to carry on an expanding business. In a capitalistic economy, electric service was, as a matter of course, undertaken for profit, and although some cities set up their own municipal systems, privately owned companies predominated. These corporations held exclusive franchises to supply what was coming to be regarded not as a luxury but rather an essential of life.

In the late nineteenth century, indeed, concentration had replaced competition throughout so much of the economy that amid exultation at the marvels of technology and outpouring of material goods, there arose social and political movements demanding popular control over the great combinations or "trusts." The quasi-radical Populist Party went down to defeat in the 1890s, but in the early twentieth century, millions of Americans became uneasily aware that the headlong exploitation of material resources as well as human labor had entailed waste, confusion, and social injustice. Although the Socialist Party gained adherents, moderate reformers known as Progressives dominated the political scene. Never a coherent entity, the Progressive "movement" consisted of diverse individuals and groups that focused their attention on a variety of social and economic evils for which they proposed a variety of remedies. They shared, however, a belief that "the people" should regain control of affairs and government—local, state, or federal—should initiate action for "the public interest."

A number of Progressives concentrated on the question of how to realize the full potential of electric technology for the benefit of everyone. Among these reformers, Morris L. Cooke of Philadelphia played a prominent part. Cooke, a mechanical engineer, was also a disciple of Frederick W. Taylor, the founder of scientific management, which applied systematic formulas to examine and organize the production, sales, and administrative functions of a business. Cooke viewed this system as the key to an unimaginable abundance that would enable human beings to create a more harmonious and spiritual civilization. Since he possessed both wealth and a sense of public responsibility, he devoted much of his time, both as a private citizen and often an appointed official, to various attempts to apply the principles of planning and rationality to "the public business." Engineers, he believed, had an inspiring function in society: to liberate mankind though applying science to human needs. Like his acquaintance economist and social critic Thorstein Veblen, Cooke sharply distinguished his profession from profit-oriented enterprise and protested "against the assumption that business—big or little—is Engineering."

To the group of "power progressives," technological advance had resulted from cooperative social effort to master the material world, but a minority had seized the fruits for themselves. Progressives denounced the attempts of corporations to gain control over waterpower sites, accused them of gouging consumers through exorbitant rates, and charged that in pursuit of the largest possible profit, they skimmed the cream of the market and neglected to serve the rural population. Enthusiasts for technology, the Progressives maintained that all should be able to use electricity freely and looked forward, with Cooke, to the era predicted by electrical engineer Charles Steinmetz, when electricity "is going to be so cheap that it is not going to pay to meter it." They differed among themselves on the most effective method to hasten that day. All agreed that government must take action, but while some demanded public ownership of all electric systems, others believed that regulation by public commissions would be sufficient, and yet others proposed public ownership of certain facilities— particularly in hydropower—that could serve as "yardsticks" to measure the performance of the private industry. Essentially Cooke held the last view. Though frequently critical of the regulatory commissions, he refused to commit himself to the cause of full public ownership, but enthusiastically supported several major government power projects.

Up until the First World War, which the United States entered in April 1917, these reformers had made progress in raising the power issue for discussion, frequently in connection with the movement for the conservation of resources. They had tasted victory in certain specific cases—Cooke, for instance, as director of public works in Philadelphia, had forced the local utility company to reduce its rates somewhat—but found after the war that the wave of reform had subsided, and conservative business interests again held firm control of government and set the tone of political discussion. As Washington administrations looked on benignly at the increasing concentration of business, the flourishing electric industry especially was rapidly coming into the hands of promoters who gathered local units into pyramids of holding companies that often milked the operating companies. "The power combine," lamented the power progressive Judson King, "now is more powerful politically than the railroads."

Technology had advanced so that by 1920, engineers could transmit high-voltage current over distances of two hundred miles or more, and many claimed that huge "superpower" generating stations and long-distance transmission lines could link extensive territories into continuous networks. Yet inexpensive power remained beyond the reach of millions of people. While city dwellers grumbled at the cost of current to light their homes as well as run their toasters, vacuum cleaners, and washing machines, ordinary country people had no electricity at all. Rural families still lit their houses and barns by kerosene lamps, and carried on domestic tasks by exerting human muscle.

Though the political situation was discouraging, power progressives stuck to their purpose and doggedly fought on. J. D. Ross, head of Seattle's municipal system, warded off encroachments from private companies; King, who had adopted the cause as one in which he could "help construct one or two new steps in the stair-way of justice," lobbied in Washington for the National Popular Government League. U.S. senator George W. Norris of Nebraska led a campaign to have the federal government develop hydropower at an unfinished wartime plant in Muscle Shoals, Alabama, on the Tennessee River. Other progressive (or "liberal") politicians, such as Franklin D. Roosevelt of New York and Gifford Pinchot, a combative conservationist and former chief forester of the United States, also worked to achieve state or federal control over the utility companies and make cheap power widely available.

Financially, socially, and professionally secure, Cooke in his fifties enjoyed good health and an overflowing supply of energy. In later life he told a young friend, "I advise you always to have twenty causes you are working for: nineteen will fail, but one will succeed." For years Cooke had been engaged in a running battle against utility company influence in the professional engineering societies, during which the council of his own American Society of Mechanical Engineers had officially censured him. In the early 1920s, he led a temporarily successful but short-lived revolt of the liberal engineers. An apparent refusal of the societies to permit organized discussion of electric rates led Cooke to believe that "the engineering profession in being silent in the face of the obvious facts is almost making itself a party to a conspiracy, which bears especially hard on the women of the land." He made his office a center of information and encouragement for the circle of activists in the cause, and personally initiated the financed studies of the cost of distribution of electricity to domestic consumers—that is, from the local station to the place of actual use. This was a touchy point with the companies, since they defended their domestic rates as necessitated by unavoidably high costs of distribution.

In 1922, when Pinchot won election as governor of Pennsylvania, he called on Cooke to advise him on utility questions. Cooke proposed and himself directed the Giant Power Survey Board, authorized not only to study the water and fuel resources of the state but also to recommend "such policy with respect to the generation and distribution of electric energy as will … best secure for the industries, railroads, farms, and homes of this Commonwealth an abundant and cheap supply of electric current." In its report, *Giant Power* (1925), the board presented a plan to control the corporations, to utilize Pennsylvania's coal resources, and to make full use of the latest technology. The state, it proposed, would direct a reorganization of the industry and would license "mine-mouth plants"—power-generating stations located next to the mines—in the bituminous fields of western Pennsylvania, from which "coal by wire" would radiate on long lines to industries and to rural households throughout the state. The project, Pinchot told the legislature, could lift "most of the drudgery of human life … from the shoulders of men and women who toil." A gigantic monopoly, he warned, might soon control "the greatest material blessing in human history." The people must act quickly, for "either we must control electric power, or its masters and owners will control us." But his plea fell on unheeding ears, and the plan never reached the floor for debate.

In New York State, Roosevelt became governor in 1929 and in his turn called on Cooke for counsel. His predecessor, Al Smith, had preserved state waterpower sites in public hands, and now Roosevelt called for "the utilization of this stupendous heritage." He appointed Cooke to the new Power Authority of the state of New York to plan for the disposition of vast supplies of current, which they (mistakenly) believed would soon be provided by the Saint Lawrence–Great Lakes project. At Cooke's instigation, the authority studied distribution costs. Confronted with inadequate data— for the industry gave out only sketchy information—Cooke reminded staff members that their purpose was not to reach unassailable conclusions but instead prove a case for action. "We can once and for all discredit the present system of domestic electric rates," he asserted. "A large part of our job is to inspire the people of the State generally to a realization of what cheap and abundant electricity will mean." Early in 1933, the authority sponsored a conference on domestic rates. Noting that smaller bills would encourage people to increase consumption and such heavier purchases would lower the costs per unit of electricity, authority experts reported that at the current average usage, the rates might well be lowered from the prevalent 7 to 3 or 4 cents per kilowatt hour. Tentative though they were, the figures were solid enough to impress some spokesmen of the industry.

Considering the circumstances, the power progressives made remarkable headway during the 1920s. So blatant were the holding companies, and so desirable was cheap energy, that the reformers in this field could count on widespread, if unorganized, support. Twice they gained congressional approval (frustrated by presidential vetoes) of the Muscle Shoals proposal. They initiated state and federal investigations, and developed the power question into a prominent political issue.

In the Great Depression that began in 1929, holding company empires collapsed, big business in general was discredited, and led by Roosevelt, a Democratic administration initiated the period of liberal reform known as the New Deal. It was a heady time for the power progressives. Conservatives themselves clamored for government action, while intellectuals and administrators advocated economic and social planning. The new president's record and some of his appointments, like that of Chicago progressive Harold L. Ickes as secretary of the interior, suggested that the New Deal might open a new electric era.

At the outset, Roosevelt adopted and enlarged the scope of the Norris proposal for Muscle Shoals; he wanted to plan for the whole Tennessee Valley, "tying in industry and agriculture and forestry and flood prevention, tying them all into a unified whole over a distance of a thousand miles so that we can afford better opportunities and better places for living for millions of yet unborn in the days to come." In May 1933, Congress assigned to the Tennessee Valley Authority (TVA) a mandate to develop the river and a broad responsibility for the welfare of the inhabitants of the entire Tennessee Valley. With Hoover (or Boulder) Dam already under way, the government constructed other major hydro projects, such as Bonneville and Grand Coulee, that today furnish current throughout a vast area of the Pacific Northwest. A revitalized Federal Power Commission surveyed power resources and costs, and reached conclusions similar to those of the New York Authority. This kind of pressure, often repeated on the local level, brought results in a general lowering of electric rates. Congress, moreover, tackled the organization of the industry when it passed the Public Utility Holding Company Act of 1935, which was intended to bring the corporations under more effective regulation and eliminate those holding companies that served no useful service function.

Cooke jubilantly participated in most of these activities and yet saw them as only the beginning of what should be done. More technically minded than many of his fellows, who rejected any contacts with the industry, he instead advocated "power pooling" to create regional networks of transmission lines to carry energy from sources both public and private. He was convinced, furthermore, that the national government ought to set forth the principles of a long-range policy to guide the development of energy for the future. In 1934, when the president selected him as one member of an advisory National Power Policy Committee, he looked forward to drawing up

The Guntersville Dam on the Tennessee River, built by the TVA in 1935–1939, creates a reservoir for flood regulation and provides some ninety-seven thousand kilowatts of installed generating capacity.

such a statement. Cooke was disappointed; although the group was largely responsible for the Public Utility Holding Company Act, it proved to be too deeply divided even to make headway on the broader task.

In fall 1933, Cooke was appointed to chair the Mississippi Valley Committee. Though the committee's original function was to advise on a mass of applications for federal aid to public works projects, he saw a considerable chance that it could "become the nucleus of an ultimate scheme of national planning." The committee drew up a proposal for the whole interior watershed of the United States. The members hoped to coordinate flood control, soil conservation (for which Cooke had become a passionate propagandist), hydropower and rural electrification, and even land use and the preservation of the wilderness. Later, Cooke chaired a committee on the drought-stricken Great Plains and, after the Second World War, a committee on water resources policy. All plans, he insisted, must take into account a complex of interrelated forces, and by the late 1930s, he had "discovered" the scientific field of ecology and was pleading for "total conservation."

But the administration's interest in planning, never profound, had largely evaporated after the first year or two of the New Deal. No subsequent project was charged with the responsibilities of the TVA, and that agency itself came to be chiefly a supplier of electricity. Cooke's proposals for regional and coordinated development, since they implied quite-centralized national control, aroused antibureaucratic fears and offended a host of particular interests—rival government agencies, large corporations, and local landowners and entrepreneurs. A number of features were separately adopted, but the vision of watershed planning never gained acceptance.

It was to rural America that Cooke made his greatest concrete contribution. In the 1920s, he had reproached the electric industry for its failure to serve the countryside, and by the 1930s, could point out that in a decade, the proportion of farms supplied with central station electricity had risen only from 1.6 percent in 1920 to 10.4 percent in 1930.[1] From the earliest days of the New Deal, he prodded officials as well as fired specific proposals at Secretary Ickes and the president. Others were pushing in the same direction: Senator Norris, of course; farm organizations; and TVA and pioneer cooperative groups in the territory that it supplied with current. Finally, in spring 1935 the president, who had been provided with a large fund for unemployment relief, allocated a $100 million to set up the Rural Electrification Administration (REA) and appointed Cooke as its head.

The agency began amid confusion; it was to lend money to construct distribution lines—but to whom? Private companies were willing to take the money, yet would accept no government control over the rate structure. Cooperatives emerged as a practical solution after a difficult year, and under the forceful leadership of John Carmody, who in 1937 succeeded Cooke as administrator, they multiplied rapidly. With REA assistance, rural neighbors joined together, borrowed funds at moderate interest from the REA, contracted (usually) for power from the nearest supplier, and built the lines to light their houses, pump the water, and milk the cows. The REA devised a number of improvements to cut down on costs and insisted on "area coverage" in order to form a compact body of customers in the territory of each cooperative. Low rates would encourage higher consumption of current. Ridiculing "the electrified farmer in the New Deal dell," utility companies tried to stop the REA through court actions, and in some territories attempted to forestall the cooperatives by rapidly stringing lines to the more prosperous residents—a tactic that resulted in "the battle of the shovel." As the REA nevertheless made headway, the companies began to extend their rural service in earnest. By the mid-1950s, 96 percent of rural

With the establishment of the Rural Electrification Administration in 1935, power reached into farmlands like these, where linesmen are stringing electric lines.

dwellings had access to electric energy. Those who have lived without that convenience can best understand the difference it made.

The power progressives enjoyed their greatest influence during the five or six years of the New Deal. By 1939, and the ensuing years of war, the demand for energy to run war industries overshadowed innovation or consideration of long-range policy. While the REA marked time, the TVA expanded enormously as a supplier of power. After the war, the liberals proposed a Missouri Valley Authority, but without success; Cooke chaired the president's Water Resources Policy Committee (his last public office), but President Harry Truman refused to take action on its three-volume report and recommendations for a system of planning for "ten rivers in America's future." The reform period had ended, and conservative forces again ruled almost unchallenged.

Technologically, the nation enjoyed a tremendous expansion of energy facilities. Although creations of the New Deal such as the great hydro projects and the REA had become too popular to dismantle, the power companies had recovered their position, and in the 1950s, a Republican administration hailed the "partnership" of government and industry. Networks of high-voltage transmission lines covered major areas of the country, and in the 1960s, mine-mouth plants, heralded by Cooke in the 1920s, converted the coal of southwestern deserts into power for the cities of the

West Coast. With nuclear energy on the horizon, it seemed clear to Cooke's friend Leland Olds, onetime member of the Federal Power Commission, that "low-cost electricity is an inexhaustible resource," and he urged public systems to structure their rates so as to reward consumers who used current most heavily in all-electric homes. Millions of Americans indeed took for granted the flow of electricity, whether to manufacture goods, cook their meals, or light, heat, and cool their houses.

In the 1960s, movements for social change arose again in America and reopened a wide range of issues, among them questions of energy. In the 1970s, a growing number of critics were discussing the situation in terms that in some ways echo, and in other ways contradict, the progressives of an earlier day. In language reminiscent of the 1920s and 1930s, consumers inveigh against monopolistic corporations and high utility rates; this time, however, they are more likely to find cheaper public electric service within their reach, obtainable perhaps from the TVA or New York State Power Authority. But many searching critics sound a far-different note from the power progressives, for they reject the goal of unending expansion. Resources of energy are not inexhaustible, they declare. Pointing out that the strip-mining of coal devastates the land and that mine-mouth plants pollute the desert air, they insist that nuclear plants endanger the lives of hundreds of thousands of people. Therefore, they call on Americans to reverse direction and set themselves to consume not more, but less, electric energy. They demand, indeed, a reconsideration of the purposes and direction of technology.

Though in their generation the power progressives, eager to disseminate the genuine benefits of "giant power," failed to foresee all the consequences of success, they also maintained that a society that has created the marvels of technology, can direct those achievements to enhance the well-being of both the people and their environment. With all his zeal for electric development, Cooke insisted also on the integration of the resources of the soil, rivers, and human knowledge into a policy for the nation.

Note

1. Another 3 percent of farms had their own generating plants.

23   Enrico Fermi and the Development of Nuclear Energy

Lawrence Badash

Speaking in the flush of early excitement over the nuclear reactions induced in particle accelerators, the great British scientist Ernest Rutherford cautioned that commercial utilization was virtually impossible because the energy released in these reactions was far less than that needed to run the machines producing them. "Anyone who expects a source of power from the transformation of these atoms is talking moonshine," he warned in a 1933 *New York Herald Tribune* interview.

Exploring What Holds Matter Together

The origin of the concept of atomic energy is shrouded in antiquity. Ancient Greek ideas of elements included the understanding that other materials were made of combinations of the elements. This raised the question of what held the elements together in these combinations. Aristotle's four-element theory may seem simplistic, but one interpretation of it has the earth, water, and air representing the solid, liquid, and gaseous states of matter, while fire was the symbol for energy. Thus, taking some historical liberties, we might argue that Aristotle's energy, not points or hooks, was the glue that held together the atoms of Democritus.

On much more solid ground, and much closer to the present time, we may look to the early nineteenth-century experiments in which substances were electrochemically dissociated into their constituents. This logically suggested that electric forces joined atoms. Later in the century came the great laws of thermodynamics and conservation of energy along with the kinetic theory of gases. These advances gave understanding to the different manifestations of energy, then quantified them and explained that the energy of motion of gas molecules depended on their temperature.

The discovery of radioactivity by Henri Becquerel in 1896 raised to a critical level the question of just what was occurring within the atom or molecule. For Becquerel found that uranium emitted a powerful radiation and appeared to do so continuously. Did this energy come from the motion of particles, breaking of bonds, or some other source?

In Paris in 1903, Pierre Curie and Albert Laborde found that a sample of radium maintained itself at a temperature higher than its surroundings—a most curious circumstance. Curie's statement that "he would not care to trust himself in a room with a kilogram of pure radium, as it would doubtless destroy his eyesight and burn all the skin off his body, and probably kill him," was quoted widely in the press and added to the public fascination with radium. Across the Atlantic, in Montreal, Rutherford and Frederick Soddy in 1902–1903 succeeded in explaining the phenomenon of radioactivity. Atoms of uranium, thorium, radium, and other related elements were, they said, being spontaneously transmuted into other atoms. The process involved the expulsion of some types of particles: the beta particle, which was already known to be an electron, and alpha particle, which would be identified in a few years as a charged helium atom. These chunks of matter were ejected with great velocity, and the relatively massive alpha therefore carried considerable kinetic energy. It was the loss of this energy—in collisions with neighboring atoms and molecules—that raised the substance's temperature, as found by Curie and Laborde.

If the mechanics of the energy expulsion were somewhat understood, the reason for the radioactive decay was not. This touch of mystery excited editorial writers and promised future insights into the nature of atoms. But the prodigious production of heat, a more tangible concept, became the hook on which to hang the worldwide popularity of radioactivity and especially its most energetic substance, radium. Marie Curie became "Our Lady of Radium," poems were written about her, and someone even asked permission to name a racehorse after her. Radium lamps to light bicycles were felt to be just around the corner, while pea-size fuel sources to send ships across the ocean were deemed only slightly further in the future. Radioactive materials were to provide a cornucopia of energy. Yet this seemingly unlimited source posed serious conflict with the law of conservation of energy. Where did the energy come from? Sir William Crookes suggested that radioelements selectively filter out the faster-moving air molecules and extract their excess energy. The Curies preferred to think of an undetected radiation that pervaded space, whose presence was known only by the alpha, beta, and gamma radiation it stimulated heavy elements to emit. The German team of Julius Elster and Hans Geitel as well as Rutherford and Soddy were closer to the mark in believing that the atom itself possessed the energy, which was released when the atom experienced an internal rearrangement.

But could the energy be released on demand and not just at the rate characteristic of each radioelement, its half-life? One line of research attempted to alter this decay rate, to speed it up or slow it down. Extremes of temperature and pressure along with different chemical combinations of the radioelements were tried to no avail. Another approach was a series of attempts to split the atom itself, no doubt anticipating some sort of energy change. Rutherford's former teacher in the University of Cambridge's Cavendish Laboratory, J. J. Thompson, was reportedly "so anxious to bust atoms artificially that ... he would have tried it with a cold-chisel before long."[1] But no one ever found large pieces of atoms broken off as a result of X-ray bombardment, for example. Others subjected atoms to intense electric and magnetic fields, detecting some effects that we now recognize as due to the atom's external electrons, but not irretrievably disrupting the atom itself. Yet the work done in Rutherford's Manchester laboratory by Hans Geiger and Ernest Marsden around 1910 showed that the alpha particles from naturally decaying radioelements were useful projectiles for learning about the targets they struck.

This led to Rutherford's nuclear model of the atom in 1911, and eventually to his disruption of the nitrogen nucleus by alpha bombardment in 1919, producing oxygen and hydrogen. In turn, this led to the conception and development of "atom smashers." In the 1920s, Rutherford, now in Cambridge, with James Chadwick successfully transmuted numerous elements in the lower region of the periodic table, using alphas from radioactive sources. But they recognized the need for a more copious beam of projectiles and greater energy to cause even more transmutations. John Cockcroft and E. T. S. Walton, also students of Rutherford, developed a high-voltage machine with which in 1932 they were able to bombard a lithium target with protons, producing two helium nuclei. Rutherford's work of 1919 deliberately produced nuclear transformations using projectiles from naturally decaying radioelements; the work of Cockcroft and Walton did the same using artificially accelerated projectiles. In this reaction, they also verified for the first time Albert Einstein's famous $E = mc^2$ relationship. Another outstanding event in that remarkable year of 1932 was Chadwick's discovery of the neutron. The neutron, of about the same mass as the proton, was uncharged. This meant that it was not affected by the intense electric fields surrounding the atom and could strike target nuclei more easily than other projectiles.

This then was the situation at the time Rutherford cast cold water on the idea of extracting usable energy from the atom—or more precisely, the nucleus—by calling it "moonshine." On the one hand, there were particle accelerators, the most famous of which was Ernest Lawrence's cyclotron developed at the University of California at Berkeley in the early 1930s. It was the prototype of a series of ever-larger machines in which the projectiles were electromagnetically accelerated in a circular path. Cyclotrons—and linear accelerators also—were able to throw protons, electrons, and heavier particles at their targets, but although some of the reactions released sizable amounts of energy, the energy input was always greater than the output. The other line of development, inducing reactions with neutrons (which cannot be accelerated since they are unaffected by electromagnetic fields), similarly gave little hope for a net energy output, again because the efficiency of neutron production and neutron-induced reactions was too low. Hence Rutherford's statement, based on existing knowledge at the time, was perfectly accurate. Einstein and Robert Millikan made the same responsible prediction. It took the discovery of a new phenomenon of nature—nuclear fission—to alter the outlook.

Fermi and the Control of Nuclear Fission Reactions

Fission was discovered at the end of 1938 by Rutherford's former student, the radiochemist Otto Hahn, and his colleague Fritz Strassman in Germany. It was explained as the rupture of a uranium atom into two roughly equal pieces when struck by a neutron. The scientific community was fascinated with the news for a number of reasons. First, it was so unexpected, with physicists and chemists having previously failed for decades to find atoms breaking off pieces larger than alpha particles. Second, the amount of energy released in the fission of a uranium nucleus was enormous—about two hundred million electron volts. When this is compared with the one or two electron volts per atom released in the most violent chemical reactions, the significance of nuclear reactions is understandable. And third, there was the strong likelihood that in the fission process, not only were two fragments produced, but extra neutrons might be released as well. If one of these extra neutrons struck another uranium nucleus, causing it to fission, and it in turn released neutrons, one of which hit yet another uranium nucleus, and so on, a sustained chained reaction was possible. If the chain reaction were controlled, we would have a continuous source of energy. If more than one neutron were released in fission, and most of them successfully split uranium nuclei, we would have a geometric increase in the number of fissions. In fact, the process occurs so rapidly—the energy is released so quickly—that we have a bomb.

These facts were recognized by physicists the world over. On no one did they make more of an impression than Enrico Fermi. At an early age, Fermi was recognized as Italy's outstanding theoretical physicist, and his colleagues looked to him to restore Italy to its place of honor in the world of science—lost since the days of Galileo, Luigi Galvani, and Alessandro Volta. An uncommon but thoroughly successful switch to experimental physics followed from brilliant work on beta particle decay and the creation of "Fermi statistics." During the mid-1930s, Fermi examined neutron-induced reactions extensively, and discovered that unlike the case with charged projectiles, *slow* neutrons are more effective in causing reactions than those moving with high velocity.

Fermi had used the occasion of the Nobel Prize awarded to him in 1938 to leave Benito Mussolini's Italy in order to settle in the United States. First at Columbia University and then at the University of Chicago, Fermi, Leo Szilard, and others showed that certain materials, such as extremely pure carbon and heavy water, were suitable for slowing the neutrons but not absorbing them. Since heavy water was virtually unobtainable, graphite bricks were used to build larger and larger piles, or reactors. Distributed among the bricks were lumps of uranium metal and uranium oxide; the fission occurred within the heavy element, and the released neutrons lost much of their energy while passing through the carbon moderator before striking other uranium nuclei. On December 2, 1942, in Chicago, Fermi and his associates achieved the first successful controlled release of nuclear energy. Cadmium rods, which absorbed many neutrons, were pulled out of the pile to allow the neutron flux to rise; they were then adjusted to control the neutron level, and reinserted to slow or stop the reaction.

Enrico Fermi works in the physics laboratory at the University of Chicago, where he and his associates achieved the first controlled release of nuclear energy on December 2, 1942.

But what had this work of a basic and applied scientific nature to do with world events? By this time, of course, the United States had joined the Allies in World War II, and like them was quite fearful that Adolf Hitler was making an atomic bomb. This danger had stimulated scientists to present the possible development of nuclear weapons to the American and British governments, and the Manhattan Project was the most significant result. Fermi's success showed three things. For one, a chain reaction was possible, and this meant that a bomb might be made (which is not to say there were no hurdles to overcome here). Moreover, since the part of

The first self-sustaining nuclear reactor, partially completed in this photograph, was a crude assembly of graphite bricks and uranium built under the stands of a University of Chicago athletic field.

uranium that fissioned was an isotope (U-235) that existed in the natural metal to the extent of only 0.7 percent and could not be separated from it by chemical means, an entirely new element, not found on earth, could likely be produced in a reactor, and this new element would be fissionable. The new element—plutonium—would be made by bombarding the otherwise-wasted isotope of uranium (U-238) with neutrons and be separable by chemical techniques. The third understanding was that reactors could also be used to produce energy.

U.S. Reactor Development

Plutonium production during World War II was accomplished in five huge reactors built on the banks of the Columbia River, in Hanford, Washington. While they made no use of the heat generated and were not prototypes of the energy-producing reactors of later generations, much valuable experience was gained. Major accomplishments here were minimizing the effects of radiation on structural materials, the technique of canning uranium cylinders in aluminum to prevent the cooling water from picking up radioactivity, and recognition of the means of avoiding reactor shutdown due to the inevitable production of neutron-absorbing fission fragments. Plutonium produced in the Hanford reactors was fashioned into the weapons tested in Alamogordo, New Mexico, on July 16, 1945, and exploded over Nagasaki, Japan, on August 9, 1945.

After World War II, reactor development was still largely associated with the production of fissionable material. An important decision that determined the direction of future U.S. technology was to use enriched uranium as reactor fuel. The Hanford reactors had used cylinders of natural uranium, which meant that the U-235, present to the extent of 0.7 percent, provided the neutrons that turned some of the U-238 portion into plutonium. But the Manhattan Project had perfected the means of physically separating U-235 from U-238, and indeed the bomb dropped on Hiroshima on August 6, 1945, was constructed of this lighter isotope. If, though, the uranium stock was enriched not to bomb purity of over 90 percent but only to about 2 to 3 percent U-235 content, it would supply excellent reactor fuel, with greater production of neutrons to allow the use of ordinary light water as moderator instead of less neutron-absorbing, but more expensive, carbon or heavy water.

The significance of this decision may be seen in the circumstance that however controversial the subject of reactor proliferation may be, American models sent to other countries are less likely to be diverted to production of plutonium for weapons because they must rely on the United States for their next load of enriched fuel. (This situation is changing, as other countries are developing enrichment capabilities. A country could, of course, content itself with the weapons producible from one reactor loading and not care about its future fuel supply.)

The American Atomic Energy Commission (AEC), which took office in 1947, sponsored research on a variety of reactor designs. Yet its overriding concern with nuclear weapons precluded a forceful and applied civilian reactor program. Somewhat by default, more by the Cold War pressures of the times, and mostly because of the driving personality of Admiral Hyman Rickover, the direction taken by nuclear power was that of ship propulsion rather than electricity generation. Conventional submarines had to surface or snorkel for air needed in the battery-charging cycle. A nuclear-powered submarine would not depend on air at all, and this ability to cruise the depths indefinitely was recognized by many of Rickover's colleagues as a revolution in undersea warfare. Nevertheless, Rickover alone felt that the technology was sufficiently developed to begin construction in the early 1950s, instead of requiring an endless, leisurely series of design studies. His perseverance led to the launching in 1954 of the *Nautilus*, a vessel powered by a Westinghouse pressurized-water reactor, and in 1955 the *Seawolf*, which carried a General Electric sodium-cooled power plant. Over seventy nuclear-powered submarines have subsequently been built by the United States (which now constructs no other types), and many more by other countries.

The advantage of nuclear power for submarines is clear, where avoidance of detection is a paramount concern. For surface ships the question of economics is of equal importance, and the U.S. Navy has constructed but a handful of nuclear aircraft carriers, cruisers, and frigates. The nuclear merchant vessel *Savannah* made her maiden voyage in 1962, accompanied by widespread predictions of a new era in low-cost shipping, but reluctance to welcome her into foreign ports because of radiation-leakage fears and debilitating union disputes undercut her value to such a point that within a few years, the ship was placed in mothballs. The Soviet Union has the nuclear icebreaker *Lenin* operating satisfactorily in its Arctic regions, and West Germany uses the *Otto Hahn* as an ore carrier, while the Japanese merchant ship *Mutsu* has been plagued by radiation leaks and blockaded from port by angry fishermen, such that its future is uncertain. The American attempt to power an airplane with a nuclear reactor was abandoned in the early 1960s when no one could conceive of a realistic mission for it.

Concurrent with the early submarine reactor work, the AEC began to encourage industrial interest in central-station electric power generation. But because the subject was shrouded in government security regulations, those engineers and executives who were allowed a view of it in 1952 and 1953 regarded it with uncertainty. On the one hand, nuclear power might well be economically competitive with fossil fuel plants. On the other hand, the government prohibition of the private ownership of reactors discouraged industrial initiatives.

In 1953, the new administration of President Dwight Eisenhower decided to make the sharing (with safeguards) of civilian nuclear information and materials a major facet of foreign policy. The president's "Atoms for Peace" program naturally had domestic impact, for the technology to be shared had yet to be fully developed. Before the end of the year, the first "full-scale" electric power demonstration plant was authorized to be built in Shippingport, Pennsylvania (60,000 kilowatts). In 1954, a law was passed authorizing the private ownership of reactor facilities, reserving to the government only possession of the nuclear fuel. These actions and many others showed the determination of the government to encourage industrial development of nuclear power. Since rapid success would be most stimulating to the corporate executives, the Shippingport plant used only an upgraded model of the navy's *Nautilus* reactor and was the responsibility of none other than Rickover. Because of Rickover's penchant to remain with something that worked well, along with his cancellation of sodium-cooled reactor development, the American nuclear industry has been heavily committed to water-cooled designs.

The Shippingport pressurized water reactor (PWR) was completed in 1957, four years after its authorization. In 1960, two other reactors, designed expressly for commercial electric power production, came online—one of them a PWR in Rowe, Massachusetts (Yankee, 140,000 kilowatts), and the other a boiling water reactor (BWR) in Morris, Illinois (Dresden, 180,000 kilowatts). Operating experience from this first generation of about half a dozen reactors proved the technological feasibility of this energy source; in fact, they worked better than expected. The costs, however, were well above those for fossil fuel plants, with construction expenses higher than anticipated, and the future clouded by a decrease in the price for coal.

Additional problems confronted utilities in the form of public opposition in the early 1960s. A reactor on Bodega Bay in northern California, which was on a geologic fault, was canceled after an earthquake interrupted construction, while a location within the metropolitan New York area was vetoed by the fearful citizens of that city. Yet in less densely populated areas, and especially in those regions with high fossil fuel costs, nuclear technology still seemed to make economic sense while the dangers of accident were relatively remote.

From 1963 on, a number of utilities decided to "go nuclear," counting on economies of scale, and encouraged by continuing financial incentives from the federal government as well as low bids by reactor manufacturers interested at this time more in creating an industry than in immediate profits. Though conceived before this period, the forerunner of the second generation of significantly larger plants was built in San Onofre in southern California. Rated at 428,000 kilowatts, and completed in 1967, its construction signaled not only the arrival of "large-scale" units but also the end of a seven-year period in which no construction licenses for commercial nuclear plants had been issued in the United States.

By the mid-1960s, nearly half of all power plant capacity ordered by American utilities was nuclear, with sizes jumping from 500,000 to 800,000 kilowatts to 1,100,000 kilowatts, and with all units but one of the PWR and BWR varieties. This commitment coincided with rising coal prices and growing public concern over air pollution from coal-fired plants. It also meshed with increasing comprehension of the concept of a "spaceship earth" with finite resources, stimulated no doubt by the American and Soviet space programs, such that at least one of the fuels of the future clearly was uranium. Nuclear projections for 1980 or the end of the century increased rapidly each year.

Many of the plants ordered since the mid-1960s are now completed. There were 58 nuclear reactors (not including military, experimental, and prototypes) in the United States as of 1976, generating about 9 percent (40 million kilowatts) of the nation's electric power. (For comparison, 20 other countries had a total of about 100 units.) More than 150 other commercial reactors have been planned or are under construction in America, with an additional capacity of some 170 million kilowatts.[2]

Nuclear Energy: Pro and Con

The future of reactors is far from clear. While some environmentalists laud their contributions to air pollution control, especially in contrast to the emissions of coal-burning power plants, and while the 1973 oil embargo by the Organization of Petroleum Exporting Countries countries encouraged the American government to strive for "energy independence," there is much controversy over whether or not reactors are a viable solution.

Utilities, faced with ever-increasing nuclear plant construction costs and expensive time delays due to inevitable litigation over siting, safety plans, environmental impact reports, and so on, and recognizing a lower growth rate than anticipated for electricity consumption, are having second thoughts about the need to go nuclear, and a number of plants have been canceled. Even more enervating to the industry than economics, however, has been the opposition of large numbers of citizens who urge legislative action against nuclear plant construction or operation, and obviously create a climate of uncertainty unfavorable for investment.

The opposition has questioned the need for fission reactors, stressing fusion, coal, oil, geothermal, tidal, wind, solar, and conservation alternatives, and has also claimed that uranium ore reserves will not last much beyond the year 2000. Industry and government data are viewed as self-serving, with the former AEC (now split into the Nuclear Regulatory Commission and Department of Energy) especially distrusted as both promoter and regulator of the industry. Above all, safety is the issue. A malfunction in the current variety of reactors cannot result in a nuclear explosion because a critical mass of required purity cannot form. Yet should a plumbing leak interrupt the flow of cooling water and the emergency system fail to function properly, there is a small but finite possibility that the reactor core will melt, a steam explosion will rupture the containment vessel, and the core's intense radioactivity will be vented and carried downwind. Similarly, an earthquake, airplane crash, or sabotage could release massive amounts of radioactivity. Should breeder reactors be brought to a level of technological and economic viability, and this is yet uncertain, it is possible that an explosive critical mass might accidentally be formed in one. While the probabilities of such accidents are extremely small, their consequences, involving many thousands of people and billions of dollars, are vast. Another cause of concern is the requirement that waste products be stored for many thousands of years without leaking into the water or air. Here the technology seems closer to realization than the identification of geologically inactive sites. Transportation of "hot" wastes and irradiated fuel rods for reprocessing, by rail or truck, is also viewed with alarm.

Accidents of this sort are but one type of danger seen by the opponents. Of greater credibility and immediacy is the creation of a so-called plutonium economy. Present reactors manufacture some of this element in their normal operation and then consume a portion; breeders would be designed specifically to make plutonium. Despite the hazards of high radioactivity, plutonium's carcinogenic properties, and the expense involved, governments or other groups willing to go to the trouble can chemically separate the plutonium from uranium, fission fragments, and so forth, and accumulate a stock of atomic bomb material. It takes only about five kilograms for a Nagasaki-size weapon, and there will be millions of kilograms of this element formed in reactors around the world by the end of the century.[3] Whether produced as part of national policy or by terrorist organizations, the proliferation of reactors is taken as tantamount to the proliferation of nuclear weapons. Regardless of international treaties requiring inspection or reprocessing in certain facilities, and despite the best protective efforts, some critics feel that the potential for abrogation of treaties, surreptitious diversion of the 1 percent of special nuclear materials expected to be "lost" in the system's bookkeeping, or outright theft makes the construction and use of atomic bombs inevitable.

Conclusion

For those who note the increasingly fast pace of life in the twentieth century, nuclear reactors provide an appropriate touchstone. Thanks largely to Fermi's genius, the first such device "went critical" just four years after discovery of the physical phenomenon on which it was based. Within three more years, reactors had been used to manufacture enough of a new element to use that substance in the construction of an awesome new weapon. Only a decade after the end of World War II, and without the intense and massive national commitment that characterized the Manhattan Project, reactors were powering vessels at sea. And during the next two decades (roughly 1955–1975), civilian reactors went from an enthusiastic vision of a limitless, safe, cheap, and pollution-free source of electric energy to one regarded as quite likely the opposite. Its advocates recognize the liabilities but believe that no technology is without risks, and that these particular dangers have been and will be well contained. They argue that the health and environmental costs of alternative energy sources are higher, and that the energy is needed for a thriving economy. Reactor opponents assume that disaster in one form or another is inevitable, and that this is a technology we should do without.

Beneath the plethora of technical data both sides produce to support their arguments, there is a value judgment to be made. For this is indeed a major social question, and the public is being given (or is taking) a rare opportunity to affect its future. Do the risks outweigh the benefits? What implications would rejection of reactors have on energy consumption, the economy, the need to control population (if resources are insufficient), or the quality of life? Is the technological imperative—if it can be done, it will be done—a desirable philosophy for the United States as it enters its third century, or is a newer attitude—a willingness to do with less—more appropriate in an increasingly interdependent world? Such questions are by no means limited to the controversy over nuclear energy, but reactors do provide, all in a single lifetime, a panorama of the successes, failures, uncertainties, and social impact of the introduction of a new technology.

Fermi saw little of this development. With the same foresight and courage that led him to abandon atomic physics for nuclear physics in the early 1930s—because he considered that the newly devised quantum mechanics would solve all the former subject's main questions—after World War II, he left nuclear physics for high-energy particle physics, feeling that this would be the exciting region for future investigation. His death in 1954 left this for others to explore, but superb teacher that he was, his students have been prominent in this next generation.

Notes

1. Henry Bumstead to Rutherford, September 30, 1905.

2. *Editor's note*: These plans met at least a temporary setback in 1979 with the accident at the Three Mile Island nuclear plant in Pennsylvania, causing widespread public concern, and resulting in a governmental study of the engineering and procedural safety precautions in nuclear power installations.

3. This estimate is found in David R. Inglis, *Nuclear Energy: Its Physics and Its Social Challenge* (Reading, MA; Addison-Wesley, 1973), 336.

24   Robert H. Goddard and the Origins of Space Flight

Barton C. Hacker

Modern rocketry began with the twentieth century in the vision of a handful of men who aspired to space travel. It flourished on the resources of governments largely preoccupied with military concerns. The achievement of space flight required both the exercise of technological imagination and marshaling of institutional support—that is, both individual creativity and social organization. Robert Hutchings Goddard, an American physicist, was one of the people whose creative vision opened the way to the age of space. His pathbreaking experiments were almost entirely his alone. With help from a few trusted assistants along with financial backing from several private or semipublic patrons, he designed, tested, and patented almost every feature of the modern rocket. But he never became part of the large-scale social enterprise that ultimately launched mankind into space. Paradoxically, the man who helped create space flight, an archetypical expression of twentieth-century corporate technology, was himself more closely allied in spirit and style to the individualistic world of the nineteenth century. Military-financed research and development programs—large, bureaucratic, amply funded projects aimed at making the rocket a reliable bomb-delivery system—proved decisive, and in these Goddard had no part.

Goddard's Career

Goddard was an experimental physicist—a profession that determined the nature of his achievement. He was born in Worcester, Massachusetts, on October 5, 1882, but spent his youth in Boston, returning with his family to Worcester in 1898. Studious but sickly, the young Goddard did not complete high school until 1904. Four years later, he graduated from Worcester Polytechnic Institute. After a year as an instructor in physics there, he began graduate studies at Clark University, also in Worcester. Goddard received his doctorate in physics from Clark in 1911, spent a research year at Princeton University, and then returned to join the Clark faculty, of which he remained a member until his resignation in 1943. By the time he was appointed full professor in 1920, his rocket researches were already well under way.

Goddard first started thinking about chemical rockets as a means of crossing interplanetary space in 1909. Notebook entries from January 24 to February 2 showed his first efforts to calculate rocket action. Some kind of self-contained reaction device based on Isaac Newton's third law—for every action there is an equal and opposite reaction—offered the only feasible method of traversing the nearly total vacuum beyond the earth's atmosphere. Chemical rockets met these requirements, although in their existing form they were little more than hollow tubes closed at one end and packed with black powder. Goddard's experiments with such rockets during 1915 and 1916 showed them to be hopelessly inefficient, and confirmed what he already suspected: better means of controlling combustion as well as better combinations of fuel and oxidizer were needed. His two basic patents for "Rocket Apparatus," issued in July 1914, had specified the use of a combustion chamber and nozzle, and suggested the potential of liquid propellants, which in theory, promised much higher energies than solids.

Goddard had gone about as far as he could on his own and was looking for outside support. He found it in the Smithsonian Institution, which in January 1917 granted him $5,000 to further his rocket researches. He scarcely had the money in the bank, however, before the United States entered the First World War. Goddard went to work for the U.S. Army Signal Corps. During 1918, he developed a solid-propellant rocket launched from a hollow tube, intended to serve foot soldiers as portable artillery. This rocket was demonstrated at the Aberdeen Proving Ground in Maryland on November 6, 1918, just five days before the armistice abruptly halted further work.

Goddard still had the Smithsonian money, though, and Clark University also provided some research funds. By 1921, he was concentrating his efforts on a rocket using liquid propellants. On March 16, 1926, Goddard achieved the world's first liquid-propelled rocket flight at his Aunt Effie's farm in Auburn, Massachusetts. More work led to further flights, notably

Robert H. Goddard stands beside the launching platform for the liquid oxygen rocket
he built in 1926 in Auburn, Massachusetts.

one in July 1929, whose spectacular brightness and noise caused a sensation. Reporters and curiosity seekers forced Goddard to look for a new site for his experiments, which he found in the nearby Camp Devens military reservation.

A more important result was the publicity that brought Goddard's work to the notice of Charles Lindbergh, the world-famous aviator as well as adviser of the wealthy and influential. Lindbergh's concerns about the future of flight after the air had been mastered seemed to be answered in Goddard's projected conquest of space. A long and fruitful relationship began in November 1929, when they met for the first time. Lindbergh quickly arranged for Goddard to see several potential sponsors, which led first to a grant from the Carnegie Institution of Washington, D.C., and then to a much larger subsidy from Daniel Guggenheim of New York. Goddard, who had been working on rockets in his spare time, could now set up a full-time experimental program in Roswell, New Mexico, starting in mid-1930.

Between 1930 and 1941, Goddard and a small group of assistants conducted over a hundred proving-stand or static tests, and attempted forty-eight rocket launches, of which thirty-one produced some kind of flight. The rockets became steadily larger and more sophisticated, reaching dimensions of nearly twenty-two feet in length, eighteen inches in diameter, and one-quarter ton in weight fully loaded with gasoline and liquid oxygen. But Goddard received no further support. During World War II, he worked on rocket-assisted takeoff and variable-thrust rocket engines for the U.S. Navy and Army Air Forces. He died on August 10, 1945, just as the war was ending. His dream of space travel was unrealized, but the large German V-2 missiles that appeared late in the war made the prospect imminent.

Goddard was by no means an obscure figure in his lifetime, but his fame has grown posthumously. Only in 1960 did the United States belatedly acknowledge his priority, in response to an administrative claim filed by Esther Goddard and the Guggenheim Foundation in 1951. The military services were using Goddard patents in missiles then under development. The government awarded a $1,000,000 settlement "for rights to use over 200 of Dr. Goddard's patents which cover basic inventions in the field of rockets, guided missiles and space exploration."[1]

The Dream of Space Travel

Despite his reticence on the subject, Goddard shared with other pioneers of rocketry the dream of space travel. Whether for reasons of personal shyness and sensitivity, the professional caution of the experimental scientist, or patriotic concern for a potentially valuable military secret, he refrained from pronouncements on his work and its possible applications. The one exception was meteorologic research, the rocket's role as a tool for studying the upper atmosphere.

Goddard had supported his request for Smithsonian money in 1916 with a substantial manuscript that he had been working on and adding to since 1913. Eventually published as "A Method of Reaching Extreme Altitudes" in the *Smithsonian Miscellaneous Collections* for 1919, the paper offered a comprehensive statement of rocket theory and rationale for development based on the value of high-altitude research. Only a modest proposal to explode a rocket-delivered load of magnesium powder on the moon and observe the flash from earth hinted at Goddard's deepest motive: the achievement of space travel. In private, however, he worked hard to build a case for space travel, especially in the confidential reports he filed during the 1920s with the Smithsonian and Clark University.

Goddard's fascination with space travel was first aroused when he encountered science fiction as a boy. The catalyst was H. G. Wells, whose *War of the Worlds* began appearing serially in the *Boston Post* in January 1898, to be followed by Garrett P. Serviss's *Edison's Conquest of Mars*.[2] The young Goddard was enthralled. Years later, in compiling material for an autobiography, Goddard recalled how the novels "gripped my imagination tremendously. Wells's wonderfully true psychology made the thing very vivid, and possible ways and means of accomplishing the physical marvels set forth kept me busy thinking." In 1932, he sent Wells a letter of appreciation for *War of the Worlds*. "I was sixteen years old," he wrote, "and the new viewpoints of scientific applications, as well as the compelling realism of the thing, made a deep impression. The spell was complete about a year afterward, and I decided that what might conservatively be called 'high-altitude research' was the most fascinating problem in existence. The spell did not break, and I took up physics as a profession."

What Goddard referred to as "completing the spell" happened on October 19, 1899. He climbed a cherry tree behind the barn. "It was one of the quiet, colorful afternoons of sheer beauty which we have in October in New England," Goddard recalled in his autobiographical notes, "and as I looked toward the fields at the east, I imagined how wonderful it would be to make some device which had even the *possibility* of ascending to Mars, and how it would look on a small scale, if sent up from the meadow at my feet.... I was a different boy when I descended the tree from when I ascended, for existence at last seemed very purposive." For the rest of his life, Goddard celebrated October 19 as "Anniversary Day" and visited the tree if he could.

But what was science fiction, and why should it have had so forceful an impact on a boy? The term "science fiction" was not coined until the 1920s, although the brand of literature it labeled was by then a century old, its first landmark being Mary Shelley's *Frankenstein* in 1818. The genre flourished during the closing decades of the nineteenth century in the works of Jules Verne, Wells, Kurd Lasswitz, and a host of less famous writers. The term itself is something of a misnomer, since technology rather than science has ordinarily been its clearest focus. Science fiction was very much a product of the Industrial Revolution: a literary form that expressed the hopes and anxieties of people caught up in the throes of rapid technological and social change. Science fiction may, in fact, be termed the fantasy of technological change, a cultural device for coping with the unsettling experience of powerlessness in the face of a too quickly altering human environment.

A major focus of that unease during the last three decades of the nineteenth century was the threat of war, which was all the more disturbing because proliferating military and naval technology rendered the prospects so uncertain. The nature of future war, especially as influenced by new technology, became not merely a topic of professional interest to military officers. From the 1870s on, first in England, and later on the Continent and in America, popular newspapers and magazines fed a growing public concern with lurid stories of invading enemy armies and the havoc they wreaked. In writing *War of the Worlds*, Wells united these concerns with scientific speculation about life on Mars and technological ideas about crossing interplanetary space, and created one of the enduring themes of modern science fiction. Wells's skill and insight made his novel something a good deal more significant than the potboilers that typified the war-scare literature, but he knew those works and used them.

Military concerns have been a potent stimulus to the technological imagination in Western civilization for centuries. Leonardo da Vinci's drawings of fanciful weapons are famous, but efforts to imagine a new technology of war can be traced back at least to Guido da Vigevano's work in the thirteenth century, and perhaps much earlier. Such speculations became heady stuff wrapped in the kind of compelling story Wells provided, and Goddard was not the only boy who succumbed. His testimony could be matched by every major figure in the early history of rocketry, and many later ones as well. Wernher von Braun, for example, recalled how he and other German youths "devoured" Kurd Lasswitz's 1897 novel *Auf zwei Planeten* "with curiosity and excitement," and Willy Ley noted how that novel had "preconditioned" a whole generation of Germans, himself included, "to taking space-travel theory seriously." *Auf zwei Planeten*, perhaps needless to say, also featured interplanetary war.

Science fiction became a bridge between realms of theory and the matter-of-fact world. Rockets may have been the only feasible basis for space travel, but at the opening of the twentieth century it took a real leap of the imagination to see the potential for journeys between the planets in the erratic black powder devices known chiefly as playthings and fireworks. Science fiction spurred that leap. It set Goddard to thinking about space travel, and his work on rockets grew directly from his efforts to imagine how space travel might be achieved.

Institutional Support and Military Concerns

Between the dream of space flight and its realization was the hard as well as expensive task of developing a reliable and efficient rocket. Turning ideas into machines took money and organization—more than any individual was likely to have at his disposal. It certainly took more than Goddard had. How to find the resources needed to reach his self-chosen goal was Goddard's constant concern. In some ways he was surprisingly successful, given the widespread skepticism about space travel that prevailed during his lifetime; in dealing with individuals, he often proved to be quite persuasive. Winning the support of large, bureaucratic organizations in government or industry, however, was not so easy for Goddard, and he endured a seemingly endless series of frustrations.

None of this was of much concern at the outset. Theorizing about space travel and rockets was cheap, and even the first rocket experiments demanded no great resources. By September 1916, however, when he first requested Smithsonian support, Goddard was feeling the pinch. "I have reached the limit of the work I can do single-handed," he explained, "both because of expense, and also because further work will require more than one man's time." Only three people were involved in the Smithsonian's decision, and Goddard got what he asked for.

Goddard enjoyed a good deal of success in winning a certain kind of institutional support. Such sponsors as Clark University, the Smithsonian Institution, the Carnegie Institution, and the Guggenheim Foundation practiced an updated version of an old tradition of private or semipublic giving. From such patrons, Goddard obtained the few thousands of dollars a year he needed to continue his experiments. Over the course of a quarter-century, the sums involved were far from trivial. Grants to Goddard from all nonmilitary sources between 1917 and 1941 totaled $209,940, of which 90 percent came from the Guggenheims. The editors of Goddard's *Papers* judge this total to be larger "than was received by any other single scientist for a single project up to the time of World War II."

The problem Goddard faced, and never really resolved, was that the kind of patrons he found could not guarantee the long-term commitment required for a full-fledged development program. Despite the large total he received over his career, his funding was always on a year-to-year basis with no assurance that money for the next year's work could be found. Almost every spring saw a flurry of letters as Goddard sought to learn if his current patron would provide for the coming academic year. Such uncertainties precluded the kind of long-term financial commitments that might have been required, for instance, to construct substantial test facilities. On this basis, full-scale rocket development was impossible since it required much greater and more reliable resources.

Organized research and development sponsored by government and industry had emerged as a significant phenomenon during the last half of the nineteenth century, shaping what has been most characteristic of modern technology. Although the individual inventor has remained important, and a surprisingly large number of technological innovations have originated in basement laboratories or their equivalent, development and production have normally demanded large-scale organization. Goddard never managed to find that kind of support, although it was not for lack of trying.

Given the obvious military promise of rockets, the military establishment and related industries seemed logical backers for Goddard's work, but he suffered repeated rebuffs. His first effort to find outside support for his research, in fact, had taken the form of a letter to the secretary of the navy on July 25, 1914, proposing the development of a rocket-propelled aerial torpedo based on his just-issued patents. The offer was rejected. With help from his friends at the Smithsonian in the crisis atmosphere of World War I, Goddard did get an army contract, but working on a small solid-propellant rocket was really a diversion from the main task. Despite the drying up of military money after the armistice, Goddard served as part-time consultant on several rocket projects for the navy and army from 1920 through 1923, none of which produced practical results; in any case, all involved solid-propellant rockets, and Goddard had decided that liquid-propelled rockets were the best hope.

Military concerns remained in abeyance through the 1920s, but revived and intensified in the following decade. As early as 1931, Goddard cited "national defense" to justify his request for the reissue of his basic 1914 patents; the request was denied. In 1933, when he was casting about for other sources of money during a hiatus in Guggenheim support, Goddard again approached the navy, without success. By 1937, rumors were leaking out of Germany of secret work on military rockets—a threat that worried both Goddard and Lindbergh. Contacts between Goddard and the military establishment steadily increased, but his quest for military sponsors remained fruitless. He had no better luck, even with Lindbergh's aid, when he approached companies like DuPont. Although some individuals were intrigued by Goddard's ideas, their organizations would not be converted to Goddard's cause.

Goddard voiced his long-standing frustration to Lindbergh in 1938. Although "the first practical use for liquid-propelled rockets is likely to be for military purposes," he wrote in a letter on July 23, the military authorities seemed unaccountably reluctant to trust civilian researchers. He noted something that scholars have found often to characterize military responses to new weapons: a resistance to change based on nothing more than habitual ways of doing things.

There is strong skepticism among military men regarding such a new development as this. For example, I recently read a manuscript by an Army officer who looks with disfavor on liquid-propelled rockets because of the difficulty of transporting liquid oxygen, forgetting the objections that were raised in the days of black powder to the thermostatically controlled magazines on board ship for modern smokeless powder.

Goddard may have been too harsh. The military officials who rejected his proposals were acting in good faith on the data they had; few of them shared Goddard's vision of the future. His rockets were costly, complex, handcrafted machines, longer on promise than performance. He could argue that they would improve, and they did, but fully a third of the test flights attempted in New Mexico were failures, and some of the successes were marginal. Even the largest of Goddard's rockets were small for military purposes, with their payloads limited to a few pounds. Expense was an important factor during the interwar years, when military budgets could seldom be stretched to meet all the needs for proven weapons, much less for a risky investment like rockets. In turning down Goddard's proposal in 1933, the navy noted that rocket devices were impractically expensive and

Goddard inspects a more advanced rocket in his Roswell, New Mexico, laboratory.

it would most likely take a long time before their performance matched their promise. That later events were to make cost seem irrelevant and less-than-perfect performance adequate does not impugn the judgment of men who did not, after all, have crystal balls, or the special incentive that promoted German rocket development—rockets were not among the weapons prohibited to Germany by the Treaty of Versailles after World War I. "The United States," as Goddard himself later observed, "had no need for long-range rockets at that time."

When World War II broke out in Europe, Goddard again sought military backing and again was rebuffed. At a major conference in May 1940, his desire to develop large rockets missiles was ignored. U.S. Navy, Army, and Air Corps officials wanted nothing more than a rocket device to assist aircraft in taking off. This was the task that Goddard undertook in late 1941 under concurrent contracts with the navy and army air forces. It was an important job, as was the work Goddard did on the development of a variable-thrust rocket engine later in the war, but it was far less than he might have done.

As far as Goddard was concerned, the German V-2 missiles that appeared late in World War II were little more than an enlarged version of his own prewar rockets. Although he was a sensitive and humane man, he was thinking more about the venture into space that such large rockets brought closer than he was about the 2,754 Londoners who perished under the bombs those rockets carried. The fear Goddard had shared with Lindbergh about German military rocketry proved to have been well founded. In 1930, the Germany Army initiated a secret project to develop liquid-propellant rocket engines. Although then modest in staff and funding by later standards, the German effort soon surpassed Goddard's lonely and methodical experimental program. Goddard may have been the first to pass each milestone, but the practical development and production of large liquid-propelled rockets were the achievements of the German team. And those achievements, rather than Goddard's, led most directly to the actual accomplishment of space flight before the next decade was over. For all his efforts to make it something more, Goddard's work remained an essentially personal project, and it died with him.

Goddard's Achievement

In evaluating Goddard's contribution to the rise of modern rocketry and origin of the space age, a clear distinction must be made between invention and theoretical-experimental work, on the one hand, and development and practical application, on the other. Goddard was one of the founders of modern astronautics, playing a vital role in formulating the theoretical basis for space flight, and devoted his life to creating a liquid-propelled rocket, which he recognized as the means to that end.

In the long run, however, his methodical, one-step-at-a-time approach and maintenance of rocket work as a personal project veiled in secrecy allowed others to outstrip him. How much of this simply reflected the realities of Goddard's situation, and how much was dictated by his character and personality, cannot easily be answered. Whether he was unable or merely unwilling to build an organization to carry on his work, he pursued a self-directed course that finally left him a lonely and isolated figure.

Salient though military concerns were in the development of rocketry, alive as Goddard was to the military role of rockets, the potential mutuality of interest never became a basis for systematic long-term support. Despite his efforts, Goddard never enjoyed the ample facilities and adequate staffing that allowed his military-backed German rivals to accomplish what he felt he could have done, given the same chance. Western civilization's unique preoccupation with military institutions has meant that military concerns have often furthered enterprise of no obvious military interest. Not only was there a significant military component in the science fiction that first stimulated Goddard and others to work on rockets but the German expertise acquired in the V-2 development program also became the starting point for space flight after World War II. Modified German military technology provided the rockets that launched the first satellites. Only then did Goddard's priority receive a belated acknowledgment.

Elaborate social organization is perhaps more characteristic of engineering than science, and the achievement of space flight is more an engineering than a scientific problem. Goddard the physicist, inspired by his personal vision of space travel, could play a part in building the theory and solving some of the basic problems, but ultimately his work had little direct role in engineering the achievement of space flight itself. That was largely a by-product of the military development of weapons of mass destruction.

Notes

1. Goddard was issued 48 patents during his life, with another 35—in process when he died—granted posthumously. The remaining 131 were issued still later, based on Esther Goddard's gleanings from her husband's notes, sketches, and photos.

2. The newspaper version of Wells's work, which appeared daily from January 6, had been altered to make Boston the scene of invasion and retitled "Fighters from Mars, or, the War of the Worlds, in and near Boston."

25 Rachel Carson and the Challenge of Greening Technology

Adam Rome

In the decades after World War II, many Americans imagined that modern technology finally would free humanity from the constraints and burdens of nature. We would overcome disease, moderate the extremes of climate, travel great distances in a flash, and enjoy abundance of all kinds. Detergents would get clothes cleaner than clean. Nuclear fission would generate electricity too cheap to meter. Plastics—seemingly inexhaustible and infinitely malleable—would end our dependence on scarce natural resources. Bulldozers would transform marshes and steep hillsides into buildable land. Soon we would live on a perfected earth, where everything was easy, comfortable, and safe.

But many of the technological miracles of the postwar period had dark sides. They polluted air and water, endangered wildlife, and disrupted vital natural processes. Some directly threatened human health. By 1970, when millions of Americans protested the degradation of the environment on the first Earth Day, the old idea of conquering nature no longer was an unquestioned good. According to one Earth Day speaker, continued technological development without equal environmental advances would lead to disaster. "What we need," another Earth Day speaker argued, "is a new technology for life."

The call to reinvent our relationship with nature has grown stronger and stronger in the years since Earth Day 1970. At different times, people have worked to further the causes of "appropriate technology," "industrial ecology," and "sustainable design." Now the drive to find greener ways to meet human needs is reshaping everything from architecture to chemistry. Often the innovators have looked to natural systems as models. Though the nation still has serious environmental problems, the new ways of thinking about technology and the environment have the potential to revolutionize society.

That revolution-in-the-making owes much to the writer Rachel Carson. Carson challenged the gung ho drive to subdue nature in a 1962 book that became one of the most influential in American history: *Silent Spring*. She indicted a marvel of the age—the new chemical pesticides that promised to eradicate a host of unwanted insects, weeds, rodents, and fungi. By unleashing an unprecedented barrage of poisons on the earth, she argued, we risked making our environment unfit for all life. But *Silent Spring* was not just a compelling and courageous critique. Carson pointed the way toward a more ecologically sophisticated approach to the problem of pests. Her call to action helped to turn a growing unease about the state of the environment into a powerful social movement. Carson also inspired many of the pioneers in the effort to develop green technologies.

Carson began work on *Silent Spring* when she was fifty, and little in her life to that point suggested that she would become a revolutionary. As a child on the semirural, semi-industrial outskirts of Pittsburgh, Pennsylvania, she loved nature and literature. When she graduated from high school in 1925, Carson planned to become a writer. Her science courses at Pennsylvania College for Women changed her mind. She majored in biology and then went to Johns Hopkins University to study zoology. But in 1934, family misfortunes led Carson to leave graduate school before she finished her PhD. She needed a well-paying job, and found one at the U.S. Fish and Wildlife Service, where she had a fifteen-year career as a writer and editor. In her spare time, she wrote for magazines. She also wrote two books about the ocean, and the second—*The Sea around Us*—was a publishing sensation. It stayed on the best-seller list for eighteen months, inspired an Oscar-winning documentary, and soon was translated into dozens of languages. A reissue of her first book, *Under the Sea-Wind*, became a best seller as well. As a result, Carson was able to quit her Fish and Wildlife job in 1952 to focus on her writing, and soon finished a third book, *The Edge of the Sea*. Though she had written government pamphlets about the importance of conserving natural resources, her popular work did not speak to issues of the day; she was instead famous as an interpreter of the wonders of nature.

In the late 1950s, however, Carson suffered an existential crisis. She always had taken comfort in the idea that human folly could not fundamentally threaten the earth. Although we might destroy forests and dam rivers, "much of Nature was forever beyond the tampering reach of man." No matter what we did, "the stream of life would flow on through time in whatever course God had appointed for it." But the technological developments of the postwar decade shattered her faith. Humans suddenly had the power "to change drastically—or even destroy—the physical world." Carson tried to "shut her mind" to the new reality. As she wrote to her closest friend in 1958, she refused to acknowledge what she couldn't avoid seeing. Yet she eventually concluded that ignoring the truth did no good. If she wanted to write anything worthwhile in the future, she needed to find a fresh way of thinking about the human place in nature.

The potential ill effects of testing nuclear weapons especially troubled Carson. At the time, test explosions were aboveground, and no one knew for sure what happened when radiation from atomic blasts rose into the atmosphere and then rained down on earth. But the media had begun to report shocking developments. Sheep died near test sites in the American Southwest, and milk had traces of radiation hundreds of miles away. So did tuna caught in the Pacific, where the United States tested new bombs on islands. Radioactive "fallout" became a public issue. Scientists and housewives organized to demand a ban on aboveground testing. A few politicians pressed the case, including Adlai Stevenson, the Democratic presidential candidate in 1952 and 1956. Religious leaders warned that fallout was not just a new form of pollution; it was blasphemy, an insult to the divine order of creation. A five-year-old boy even warned a friend not to eat snow "because there is a piece of the bomb in it."

The fallout issue led Carson to view chemical pesticides in a different light. She had worried for more than a decade that the new poisons might be problematic. In 1945, she proposed an article to the *Reader's Digest* about the first Fish and Wildlife studies of DDT's impact on birds, but she put the idea aside when the magazine turned her down. Then, just as she was struggling to come to terms with fallout, she was drawn to the subject again. In 1957, a group of Long Island, New York, residents sued to stop the government from spraying private land with DDT in order to control gypsy moths, and Carson quickly saw the importance of the suit. She also received anguished reports from several friends about the damage to wildlife from aerial spraying campaigns. She called experts to get more facts. The more she heard, the more she was convinced that the growing reliance on pesticides was "an alarming threat to human welfare, and also to the basic balance of nature on which human survival ultimately depends."

Carson's breakthrough was seeing that the heedless use of chemicals led to a neglected yet more perilous kind of fallout. Like radiation from

weapons testing, pesticides often had far-flung effects. They did not stay
in place; they persisted in the environment and could accumulate in the
bodies of living creatures. But chemical fallout did not come from just one
sector of society. Pesticides were ubiquitous. They were used by businesses,
consumers, and government agencies, and were increasingly important to
the economy. *Silent Spring* was Carson's effort to force public debate about
whether the rewards were worth the risks.

As Carson knew, the popular view of chemicals was overwhelmingly positive. Boosters and journalists argued that DDT promised total victory in the
war against insects. The "atomic bomb of pesticides" would enable farmers
to provide a superabundance of food. Homeowners could stop mosquitoes
and cockroaches with push-button ease: "Presto! No more pests!" A 1947 ad
even showed a housewife, cow, apple, potato, pet dog, and chicken singing
"DDT is good for me-e-e!" That was not just a Madison Avenue fantasy. By
the mid-1950s, one reporter wrote, pesticides were as widely appreciated as

Demonstrating the safety of DDT at Jones Beach on Long Island, New York, 1945.
[Mosquito control, Jones Beach State Park, Long Island, 1945.]
Source: Bettmann/Corbis

orange juice. Hundreds of pesticides sat on supermarket shelves, and consumers bought wallpaper and paint laced with DDT. Government officials also took for granted that pesticides served the public interest. They sprayed chemicals on residential streets, in parks and swimming pools, along roadways, and at beaches.

Though acknowledging that pesticides sometimes were necessary, Carson asserted that they were being used recklessly and arrogantly. Any human interference with the intricate interrelationships in nature might have "consequences remote in time and place," yet the boom in the use of chemicals had come "with little or no advance investigation of their effect on soil, water, wildlife, and man himself." The result was a potentially catastrophic experiment. Could we war against nature without ultimately destroying ourselves?

In Carson's view, the evidence already was deeply disturbing. Because pesticides were nonselective poisons, they killed indiscriminately. In truth, they were "biocides." The "good" species fell victim as well as the "bad"—bees as well as corn borers. Aerial spraying was particularly ill considered. The "chemical death rain" fell on birds, fish, pets, farm animals, and children. Often the harm was immediate, and Carson described in horrifying detail the convulsions and agonies of the dying. But the immediately lethal effects were only the start of the havoc. When pesticides became part of food chains, the casualties mounted. Eagles failed to reproduce after eating fish and crabs from marshes sprayed with mosquito killers, while robins died after eating earthworms that ate elm leaves with chemical residues. Pesticides also affected the ecology of human bodies; they could cause cancer, genetic damage, and fertility problems.

Carson also questioned the fundamental promise of pesticides. Were they the most effective way to deal with pests? Even judged solely by that criterion, the new chemicals frequently fell short because their design did not take into account the complexity of the living world. Insecticides especially were "self-defeating." Insects developed resistance to pesticides over time, so to keep up with insect evolution, pest controllers continually had to race to develop new chemical weapons. Many insecticides killed the natural enemies of their targets as well—and when the pests came back, as they always did, birds or other insects no longer checked them. That ensured that chemicals had to be applied again and again, at greater and greater cost.

"ANOTHER SUCH VICTORY AND I AM UNDONE."

Cartoonist Bill Mauldin takes on pesticides, 1962.
Source: *Chicago Sun-Times*, 1962. Library of Congress

Pesticides thus failed both practically and morally. Despite their seeming sophistication, they were exceptionally crude tools, "born of the Neanderthal age of biology and philosophy, when it was supposed that nature exists for the convenience of man." Yet we had a wiser way forward. Instead of using "brute force," we should "aim to guide natural processes as cautiously as possible in the desired direction." That would require a new humility—a willingness to learn from nature. The possibilities were rich. Some alternatives to chemical control were well established, Carson contended, while others were being tested or were merely imaginative ideas in the minds of scientists. We could rely on diseases that afflicted target species. We could use natural attractants and repellants that interfered with pest reproduction—or introduce sterilized males into pest populations. We also could change farming, gardening, and household practices to lessen the opportunities for pests to spread. The key in every case was a thorough understanding of biology and ecology, not chemistry.

Carson had no illusions about the obstacles to taking "the other road." She made powerful arguments about the political economy of pesticides. Unless the public insisted on new priorities, research on biological methods of controlling pests always would be slighted. Manufacturers were pouring money into universities for pesticide research, and that created attractive graduate fellowships and staff positions—but the alternative approaches never were as well endowed, "for the simple reason that they do not promise anyone the fortunes that are to be made in the chemical industry." The alternatives to pesticides also were disadvantaged because they seemed too expensive. That calculation, however, was faulty. "The true costs" of pesticides included the damage they did to the environment and human health—and those costs were prohibitive. But citizens never would get a true accounting from the authorities. Industry had a financial stake in the continued use of chemicals, and government pest controllers often were driven by a "fanatic zeal." When challenged, they offered "little tranquilizing pills of half truth," "false assurances," and "sugar coating of unpalatable facts." Because most academic researchers were funded by industry, they too could not be trusted to be honest about the hazards of pesticide use.

Ultimately, *Silent Spring* was a call for the social direction of technology. Because the tools of modern life might prove more dangerous than miraculous, the course of technological development needed to be a subject of popular debate. "It is the public that is being asked to assume the risks," Carson maintained. "The public must decide whether it wishes to continue on the present road, and it can do so only when in full possession of the facts." Her conclusion underscored that point. "The choice, after all, is ours to make," she wrote. "If, having endured much, we have at last asserted our 'right to know,' and if, knowing, we have concluded that we are being asked to take senseless and frightening risks, then we should no longer accept the counsel of those who tell us that we must fill our world with poisonous chemicals; we should look about and see what other course is open to us."

Many Americans were ready to hear Carson's warning. Though few people had the sense yet that the nation faced an environmental crisis, a variety of environmental problems began to hit home in the late 1950s and early 1960s. People in many states campaigned to save "open space" from the sprawl of suburbia. The smog over California's exploding cities became a symbol of the perils of progress, and federal health officials organized a national conference on the hazards of air pollution. Thousands of home-owners in new subdivisions watched in shock as detergent foam came out of their kitchen faucets. Many families also came face-to-face with water pollution while trying to enjoy new opportunities for outdoor recreation. In 1959, the Food and Drug Administration stopped the sale of cranber-ries before Thanksgiving because of possible pesticide contamination. Of course, fallout was a huge concern too.

Silent Spring also benefited from the vogue for social criticism. The read-ing public at the time ate up books that questioned whether the nation was sacrificing too much to the drive for material prosperity. John Ken-neth Galbraith's *The Affluent Society* (1958) contended that the new abun-dance of consumer goods obscured a worrisome decline in the provision of schools, parks, and other public goods. Vance Packard's trio of best-selling critiques—*The Hidden Persuaders* (1957), *The Status Seekers* (1959), and *The Waste Makers* (1960) —attacked conspicuous consumption, advertising, and planned obsolescence. William Whyte's *The Organization Man* (1956) derided corporate conformity. John Keats's *The Crack in the Picture Window* (1956) and *The Insolent Chariots* (1958) mocked the soullessness of the post-war suburbs and the auto industry's subordination of functionality to style. Carson drew on a few of their arguments in *Silent Spring*—and her work certainly matched theirs in challenging some of the basic institutions in American society.

The buzz about *Silent Spring* started months before the book's publica-tion. Despite a threat from a leading chemical company to sue for libel, the *New Yorker* magazine ran a three-part excerpt, and Carson's claims became the subject of heated debate. "'Silent Spring' Is Now Noisy Summer," one headline proclaimed. The Book of the Month Club made *Silent Spring* a main selection, with a first printing of 150,000 copies. In the club's newslet-ter, U.S. Supreme Court justice William O. Douglas touted the book as "the most important chronicle of this century for the human race." The Con-sumers Union—the nonprofit publisher of *Consumer Reports*—arranged to buy 40,000 copies of a special paperback edition for resale to its members. Although Carson turned down almost all requests for interviews, the few she did all became news.

The months of publicity helped *Silent Spring* to debut as the nation's number-one best seller. The trade association of the agricultural chemical industry quickly organized a $250,000 publicity campaign to discredit Carson's work. Monsanto even sent a parody of *Silent Spring*'s first chapter to every newspaper in the country. But the industry counterattack only ensured that the controversy over the book continued to be a big story. In many communities, civic organizations hosted debates about pesticides, and local media frequently covered those events. The national media also kept reporting on the issue.

By far the most important report was a prime-time CBS documentary in April 1963, *The Silent Spring of Rachel Carson*. The network estimated that ten to fifteen million people saw the show, and most had not read Carson's work or heard her speak. Unlike the men from industry and government who appeared on the program, Carson had no institutional affiliation, so CBS filmed her in the living room of her suburban Maryland home—and she more than held her own, though she was weak from treatments for cancer. She read passages from *Silent Spring*. In a calm but forceful manner, she rebutted the arguments of her critics, especially Robert White-Stevens of the chemical company American Cyanamid. For White-Stevens, Carson's biggest failure was not appreciating that modern technology was steadily gaining control of nature. "If man were to faithfully follow the teachings

Rachel Carson testifying before Congress about pesticides and environmental pollution, 1963.
Source: United Press International photo, 1963. Prints and Photographs Division, Library of Congress

of Miss Carson," he asserted, "we would return to the Dark Ages, and the insects and diseases and vermin would once again inherit the earth." Carson replied that the pesticide boosters misunderstood the human place in the world, and CBS gave her the last word. "We still talk in terms of conquest," she said. "We still haven't become mature enough to think of ourselves as only a very tiny part of a vast and incredible universe. Now I truly believe that we in this generation must come to terms with nature, and I think we're challenged, as mankind has never been challenged before, to prove our maturity and our mastery, not of nature but of ourselves."

Carson's alarm sounded at the highest levels of government. After reading the *New Yorker* articles, President John Kennedy's science adviser brought together officials from six agencies to discuss pesticide policy, and he later commissioned an official review of the subject. A month before *Silent Spring* was published, Kennedy himself acknowledged Carson's work when he was asked at a press conference whether the administration was going to examine the possibility that DDT and other chemicals had "dangerous long-range side effects." The 1963 report of the president's Science Advisory Committee on "The Use of Pesticides" was widely interpreted as a vindication of Carson's claims, and CBS broadcast another prime-time special, titled *The Verdict*, on *Silent Spring*. Congress also investigated the issue, and Carson was invited to testify before two committees. She took the opportunity to underscore her argument that pesticides were just one instance of the broader problem of environmental pollution.

Silent Spring shook up many scientists too. Though the profession had long held to an ideal of detachment from political debate, the mounting evidence of the hazards of modern technology led many scientists to become activists. The fallout issue was a catalyst for some—and many more were goaded into action by the controversy over pesticides. That was especially true in the disciplines of biology and ecology. "Rachel Carson's book *Silent Spring* created a tide of opinion which will never again allow professional ecologists to remain comfortably aloof from public responsibility," a committee of the Ecological Society of America maintained in 1965. "Ecologists have a definite obligation to make their views known when they can provide information which might avert environmental disaster." Not everyone agreed. But by the end of the 1960s, the most outspoken scientists were prophetic leaders in the growing environmental movement. As *Time* magazine wrote in 1969, they were "the new Jeremiahs."

At the grass roots, thousands of people made Carson's cause their own. Many spread the word about the dangers of pesticides by writing letters to the editor, preparing educational brochures, or sponsoring discussions. Others lobbied for restrictions on pesticide use. Some acted as individuals, but many were members of garden clubs, women's groups, or wildlife conservation societies. To Carson's surprise, the local activists frequently wrote to her about their concern, and their notes more than anything gave her the heartening sense that *Silent Spring* was "serving as a rallying cry for an awakened public."

In 1967, a lawyer in New York sued to stop his county from spraying DDT, and that lawsuit led to the formation of the Environmental Defense Fund, the first organization to join science and law in the service of environmental protection. The Environmental Defense Fund soon won a halt to DDT use in Wisconsin and Michigan. In the early 1970s, the federal government banned several of the chemicals that Carson targeted as "elixirs of death," including DDT. As Carson hoped, researchers began to give more attention to alternatives to pesticides, although the new research mostly explored the potential of "integrated pest management," which combined biological and chemical methods. By the late 1980s, the agricultural establishment was starting to consider the benefits of organic production.

The broader impact of *Silent Spring* is harder to gauge. After Carson's death, Senator Abraham Ribicoff said that her work ensured that no one in the future would consider the benefits of technology without weighing the environmental risks. "It was no longer possible," he argued, "to build a new factory without being concerned about air pollution, a nuclear reactor without being concerned about water pollution, or a new pesticide to make food more abundant without being concerned about the health of wildlife and even people." That eulogy was an exaggeration, but *Silent Spring* certainly challenged readers to think harder about the definition of "progress." By 1970, many Americans were questioning the wisdom of building a supersonic transport, boycotting nonbiodegradable detergents, and calling for a phaseout of the internal combustion engine. Congress acknowledged the new concern about the "undesirable by-products and side effects" of technological development by establishing the Office of Technology Assessment in 1972.

Carson's discussion of the road not taken also became a source of inspiration and affirmation for many people working to build a more sustainable society. She identified the need that they hoped to meet. Chemistry is the best example. Though once reviled by most chemists, Carson now is acclaimed for pointing the way to a salutary paradigm shift. As a former president of the American Chemical Society wrote on the fiftieth anniversary of the publication of *Silent Spring*, "Carson's systemic, ecological approach provides the backbone for green chemistry, which has become a major force worldwide. Now chemists choose processes that create less waste, use less water and resources, solve problems without being more powerful than necessary, and develop products that degrade and metabolize in concert with nature." That is a remarkable change.

Unfortunately Carson did not live to see many of the legacies of her work. She died of a heart attack in 1964, only a year and a half after *Silent Spring* appeared. Though she was only fifty-six, her death was not sudden. Her last five years all were shadowed by serious illness. She had radiation treatment for breast cancer that had metastasized. She also had phlebitis, ulcers, iritis, pneumonia, meningitis, shingles, pelvic fractures, and uncontrollable nausea. For a time, she feared that she would be unable to finish the book. After publication, she struggled to find the strength to continue making her case. But she did. In her last months, though still the subject of attacks from pesticide boosters, she was honored in many ways. Carson knew that she had done everything she could.

For all that *Silent Spring* accomplished, however, Carson's story does not quite have a Hollywood ending. We use more pesticides now than in 1962. Even though the controversy over *Silent Spring* pushed manufacturers to develop pesticides that are less persistent in the environment, that improvement came with a catch: many of the mainstays of the chemical arsenal today are more toxic than their predecessors. We still make many basic decisions without a full accounting of their environmental costs. But *Silent Spring* remains in print—and every day, new readers are inspired by Carson's call to become smarter and humbler in our dealings with the rest of creation.

26 Frederick E. Terman and the Rise of Silicon Valley

James C. Williams

California's Santa Clara Valley is an alluvial plain at the head of the San Francisco Bay. It is bounded on the west by the mountains of the Pacific's Coast Ranges and on the east by the Diablo Range. The two meet and pinch off the valley some thirty miles below the bay. Its northern boundary is marked by Palo Alto, the home of Stanford University, and mudflats that extend into the bay. Until the 1960s, the Santa Clara Valley was home to prune, apricot, and cherry orchards along with a world-class canning and packing industry. Today the world knows it as "Silicon Valley."

The origins of the electronics industry, which gave Silicon Valley its name, reach back a century to a time when California seemed to be almost a colony of the United States. Geographically isolated from the rest of the nation, the region developed somewhat autonomously. Its principal city, San Francisco, became devoted to shipping and eventually capital investment. Its major industrial activities were extractive in nature—mining, logging, and farming. It lagged behind the East in developing manufacturing industries partly because Californians had to import the essential industrial energy resource—coal. Yet four decades of gold mining had provided Californians with a rich knowledge of hydraulic engineering. By the 1890s, local entrepreneurs and engineers talked of harnessing waterpower to generate electricity, learning from and even using hydraulic work developed by the mining industry. "By electric transmission," observed one engineer, "[power] can be delivered to those points where it will ... enable and encourage the establishment of industries, which, without it, would have been impossible."

Regional electric power companies began building hydroelectric plants high in the Sierra Nevada mountains, using the Pelton waterwheel, developed for the mining industry in San Francisco, and generators produced by eastern manufacturers. They faced the challenge of pioneering high-tension, long-distance electric power transmission on their own if they were to bring electricity to San Francisco and other coastal cities. By 1901, they had succeeded, and in the process, power company engineers and university engineering professors developed a cooperative style of research and development.

Field experience led to invention and innovation, university laboratory experiments, and cooperative testing. For example, in 1898, Stanford professor Frederic A. C. Perrine, his students, and power company engineers successfully field-tested a regionally developed high-potential oil switch that led directly to construction of forty thousand volt lines. Perrine also took a two-year leave from Stanford to consult with the new Standard Electric Company of California, where he coordinated field and laboratory tests of aluminum wire, helping to introduce it for use in California systems. Other Stanford professors involved in similar work included E. E. Farmer and George H. Rowe.

Interchanges between practicing engineers and professors sustained high-tension, long-distance transmission advances. They shared their designs and tests of insulators, switches, and transformers, and employed scientific instruments, such as the oscillograph, to study transmission phenomena. They communicated easily, as professors seeking the opinion of "practical men" and speaking to them in an equally practical way. As one power industry representative put it, the professors tore "the mask of ambiguity from electrical theory." They did not lead us "through an interminable labyrinth of mathematics [or] ... through catacombs of theory—but instead [avoided] these dreary wastes by at once opening up a panorama of results."

Stanford University's electrical engineering program had become an important element of the California electric power industry within a decade of its opening. Its electrical engineering graduates found ready work in expanding power companies, and its professors worked closely with the industry. In 1905, Harris J. Ryan, transmission research pioneer at Cornell University, came to Stanford to head the young department. He continued the cooperative university-industry style of research and development, installing the first high-voltage laboratory in the West in 1913, and the first two-million-volt university laboratory in America in 1926.

Meanwhile, shipping interests at the growing port of San Francisco became interested in the wireless communication system developed in Europe by Guglielmo Marconi. Soon after 1900, wireless firms appeared on the Pacific Coast to serve the region with ship-to-shore communications. They also created an intense interest in radio technology among young amateurs, who built wireless sets with Quaker Oats boxes. Historian Arthur Norberg notes that "this field of wireless was ripe for innovation and excellent for exploitation with minimal capital. The only obstacle for amateur and entrepreneur alike was how to circumvent the Marconi patents."

In 1909, Cyril F. Elwell, a recent graduate from Stanford University's electrical engineering program, bypassed Marconi by purchasing exclusive American rights to Danish scientist Valdemar Poulsen's wireless patents. In Palo Alto, he demonstrated the system, and gained financial backing to organize a company from Stanford University president David Starr Jordon and several professors. Elwell's new firm headquartered itself in San Francisco and quickly expanded. By 1911, a manufacturing branch called the Federal Telegraph Company formed and established a laboratory in Palo Alto near Stanford. Now young university graduates had a backyard opportunity in a new branch of the electric field.

* * *

Into this fruitful environment in 1910 arrived Frederick E. Terman, a bright and talented boy of ten. His father, psychologist Lewis M. Terman, had joined the Stanford faculty, bringing his family from Indiana. The elder Terman already had a reputation as an expert in human intelligence and soon developed the Stanford-Binet intelligence quotient, the basis for standard IQ testing. Meanwhile, his son hunted rabbits with a twenty-two-caliber rifle in the hills of the Stanford campus, fished for bass in Felt Lake, and learned to swim in Lake Lagunita. He also caught the wireless craze, and by age sixteen, the young radio "ham" had built a transmitter with Herbert Hoover Jr.

Not surprisingly, young Terman pursued a college education at Stanford, graduating in 1920 with a bachelor's degree in chemical engineering. He worked briefly at the Federal Telegraph lab and then enrolled in Harris J. Ryan's electrical engineering program. After graduation in 1922, he was urged by Ryan and his father to attend the Massachusetts Institute of Technology (MIT). Within two years, he earned his doctorate in electrical

The Federal Telegraph Laboratory, Palo Alto, California.
Source: California History Center Foundation, Cupertino

engineering under the tutelage of Vannevar Bush and was offered a teaching position at MIT. Terman returned home in 1924 to prepare himself for his new career in the East, but he was struck down with tuberculosis.

In 1925, Terman's old mentor, Ryan, opened a small radio communications laboratory in the attic of the campus electrical engineering building. Although the young man had still not fully recovered from his illness, Ryan offered him a half-time teaching job. Terman had already made his mind up to work in the radio field and accepted at once. "I took the course in the theory of high voltage lines," he recalled years later,

and just modified it into a generalized theory of long lines. Now this included telephone lines, radio frequency transmission lines, antennas, artificial lines, filters, and so on, as well as high voltage power lines. You could generalize this theory and teach the whole thing in one package. Then the research with students increasingly got to be related to vacuum tubes and circuits associated with vacuum tubes.

Terman's experience at MIT had convinced him that Stanford was falling behind other engineering schools and needed better research with more support to catch up. In 1927, while in charge of the radio laboratory, he wrote an article in *Science* suggesting that engineering research done in corporations was better than that done at universities. He sent a copy to Ryan

Frederick E. Terman
Source: Department of Special Collections and University Archives, Stanford University Libraries

along with a letter asking for stronger university research support. "With its past reputation as a center of high voltage research, and with the establishment of the Ryan Laboratory," he wrote, "Stanford is in an excellent strategic position to initiate a pioneer movement that will make this the national research center of electrical engineering." Just a modicum of support, he thought, would do wonders for recruiting students and building prestige.

Ryan was swayed. Terman got additional support, recruited bright students, and began ambitious research projects in radio wave propagation and vacuum tube design. The radio lab soon became the focal point for what Terman described years later as "electronic nuts, those young men who show as much interest in vacuum tubes, transistors, and computers as in girls." He supervised thirty-three advanced degrees during his first six years, or half of the department's total, and after Ryan retired in 1932, more and more electrical engineering students gravitated to Terman. "Studious, soft-spoken, and forever self-effacing," wrote Michael Malone, a regional journalist, who added that Terman's ability to synthesize knowledge "made him first a brilliant teacher and later a profound visionary."

One of the things that attracted students to Terman was his outreach to the local electronics industry. He explained that he "took an interest in what companies there were because there were some jobs there and these were interesting activities anyway. I would take the boys out to see these activities, see what the world off campus was like, and sometimes have people from these companies come in to give talks to the students." This became especially important after Federal Telegraph relocated to the East in 1932, because it had been the prime local employer for early Stanford communications graduates. Federal's departure left "kind of a semi-desert out here, and one was interested in these [small] companies and helping where they could." Among the remaining firms were Heintz and Kaufmann as well as Eitel-McCullough in San Francisco, and the new Litton Engineering Laboratories in Redwood City. Both Ralph Heintz and Charles Litton were Stanford graduates.

Terman also sought linkages with other university departments. In 1929, he wrote in his annual report on the laboratory, "It is anticipated that from time to time in the future cooperation from the chemistry, physics, and mathematics department will be very desirable and perhaps necessary, in research on certain problems." Such connections were eventually made in large part because of the work of Stanford graduate William W. Hansen, who had joined the physics department in 1934. Hansen was excited by nuclear research, and decided to accelerate electrons to probe the nucleus using X-ray tubes and the Ryan High Voltage Laboratory, but a lack of money during the Depression years prevented this approach. Instead, Hansen pushed forward his research through resonant techniques and soon developed a cavity resonator to accelerate electrons, which he called a "rhumbatron."

At this time a young pilot, Sigurd Varian, persuaded his brother to open a small laboratory near their home in Halcyon, a town 225 miles south of Palo Alto. Sigurd was interested particularly in developing an aircraft navigation and detection system, and his brother, Russell, was a Stanford physics graduate who coincidentally had been Hansen's roommate when they were both students. Russell wrote Hansen with an idea that the rhumbatron might be useful in Sigurd's project. Hansen agreed, and a collaboration began that brought the brothers from Halcyon to Stanford. The university agreed to take them on as unpaid research associates with $100 a year in materials and laboratory space in return for a half interest in any resulting patents. Within a few months they had developed a new electron tube dubbed the klystron, which seemed to be the key to the navigation and detection problem. They quickly sought and received financial support from the Sperry Gyroscope Company in New York—a connection that launched increased microwave research at Stanford.

Meanwhile, Terman was named head of Stanford's Department of Electrical Engineering in 1937. Two of his students, William Hewlett and David Packard, had met on campus, and then gone separate ways after graduating in 1934. Hewlett continued with Terman for a while and then went to MIT for further study, and Packard took a job in New York with General Electric. In 1936, Terman had helped Hewlett get a job and also brought Packard back to Palo Alto with a graduate fellowship. The fellowship came indirectly from Hansen's work with the Varian brothers and Sperry Company. Litton had patented some multigrid tubes in which oscillations occurred and assigned the patents to Stanford, and they were put into the same pool as the klystron patents. Norberg explains what happened when Sperry licensed the pool of patents: "Stanford received an extra $1,000 for the Litton idea and Litton insisted the money go to the engineering school for research on electronic tubes. Terman allocated it to graduate student support, specifically for David Packard."

In 1937, Terman encouraged Hewlett and Packard to form a business to commercialize the audio-oscillator developed by Hewlett under his mentor's guidance. In their tiny shop located in Packard's garage, Hewlett perfected the audio-oscillator as a distortion analyzer that generated signals of different frequencies. The pair also produced an electronic frequency meter, and Terman was able to tell when they had orders for their devices: "If the car was in the garage there was no backlog. But if the car was parked in the driveway, business was good." By 1940, the two young engineers' growing business had nine employees and would expand rapidly during the war years.

In addition to helping Hewlett and Packard, Terman had directed several students to Hansen's klystron project. On the eve of World War II, they were successful in making the klystron a practical microwave radio device with many applications. The war brought a variety of opportunities and changes. The Sperry Company became an important war contractor, and Terman's connection to Hansen's work helped him land $10,000 from Sperry for klystron research in Stanford's communications program. Meanwhile, Sperry relocated Hansen and his physics team to its Long Island research center for the duration. Before leaving in December 1940, however, Hansen taught a special class on the klystron for Terman's graduate students, preparing them for the war work they would be doing with Sperry's support. As historians Stuart Leslie and Bruce Hevly described the impact, "Sperry's support of klystron research back at Stanford trained a new generation of microwave engineers. ... In just a few years Stanford's electrical engineering program had gone from providing a couple of graduate students for the physics department microwave studies to a full-fledged research commitment of its own."

Terman himself was affected profoundly by the war. In 1942, Bush, chair of the Office of Scientific Research and Development, asked him to move to Boston to take over the top secret Radio Research Laboratory at Harvard University. There he managed a research program involving over eight hundred people, devised jamming devices for radar, and developed tunable receivers for detecting and analyzing radar signals. He lived across the street from Harvard University's treasurer, and because Terman's project brought in over half the money that Harvard was spending on war research, the two became friendly. On Sundays, Terman used to chat with his Harvard acquaintance, following him around as he worked in his garden. "I asked him what he thought would happen after the war. It seemed to me there'd be a new wave of government support. The scientific war effort had been so very successful."

Terman's views were reinforced, and although there had been little peacetime government support of university research before the war, he became convinced this would change. "The war had made it obvious to me that science and technology are more important to national defense than masses of men. The war also showed how essential the electron was to our type of civilization." After the war "there would be, for the first time, real money available to support engineering research and graduate students. This new ballgame would be called sponsored research," he said.

* * *

Terman observed that "Stanford emerged from World War II as an under-privileged institution. It had not been significantly involved in any of the exciting engineering and scientific activities associated with the war." When he returned in 1946 and assumed leadership as dean of engineering, he moved forward with the view that Stanford could achieve a position of national significance in electronics. It already had a strong reputation in high-voltage power transmission and, he observed, "had made some very important contributions under Harris J. Ryan." But the "newer opportunities, the opportunities of the future were in electronics and things related to electronics." So Terman launched a simple plan to attract the brightest faculty who would seek out support for the most desirable research projects to attract the best graduate students. At the same time, they would develop close connections with private industry. He dubbed his approach "steeple building," and since Stanford already had an edge in microwave technology because of its klystron work, he looked to it as the foundation for the future.

"After the first month or two I was back at Stanford," recalled Terman, "a couple of boys showed up from the Navy. The Office of Naval Research [ONR] had just been established and they headed for my lab first." Terman talked over some ideas with the U.S. Navy, got university president Donald Tresidder to back him, and landed a $225,000 annual contract for basic research. "We started off with three projects. The one in chemistry fizzled. The physics project led to the Nobel prize for Dr. Felix Bloch, who discovered nuclear magnetic resonance. The project in electrical engineering was a seed that blossomed into today's nationally recognized research program in engineering." This initial start in sponsored research led to Stanford's Electronics Research Laboratories (ERL) as well as a project for the development of the Stanford Linear Accelerator Center.

The ERL carried out fundamental studies in electronics. While researchers were not doing applied research, and bristled at the idea that the ERL might become a development laboratory for military or commercial ventures, they did follow Terman's lead in building ties outside the university. After the war, the Varian brothers had opened a business in San Carlos, not far from Palo Alto, joining Hewlett-Packard, Litton Engineering, and other homegrown electronics firms. ERL researchers worked informally and sometimes formally with these and other companies converting their discoveries into practical hardware; the number of innovations coming from ERL's fundamental research gave Stanford a position of increasing importance to military and commercial interests. It is therefore not surprising that the ONR came first to Terman with an applied research proposal after the Korean War began in 1950; Terman got university support for it within two weeks.

The enormous $450,000 annual applied electronics research contract consolidated Stanford's position in electronics. Terman became director of the new Applied Electronics Laboratory housed in a new facility built with ONR dollars and a gift from Hewlett-Packard. In directing the program, he accepted only projects that strengthened Stanford's basic electronics research and enhanced the university's reputation. He also redoubled his efforts to build stronger ties with industry. Terman had been doing this for years, speaking to industrial groups about the benefits of their being acquainted with university research work and getting to know the fellows in local industries. But now this "community of interest between the University and local industry" became essential, since the Applied Electronics Laboratory was to produce prototype electronic devices and then work with the firms that actually would produce them.

A spur to this community of interest came from Stanford's own financial problems at the time. The university's endowment was too small to provide either adequate operating or capital improvement funds, and its founding grant prevented it from selling off any of its eighty-one hundred acres of land. Administrators, working after 1951 with a faculty committee, concocted several plans for generating income from the land, including agricultural development, a regional shopping center, and housing projects. A forty-acre corner was earmarked for a little light industry. Most of the planners' energy went into the shopping center and housing, and a master plan would not be finalized until 1953. But Varian Associates looked at the industrial land and took the first initiative toward developing it.

Varian Associates, according to historian Henry Lowood, specialized in developing products from research conducted at Stanford, and regularly hired university faculty, research associates, and students. It was so successful that by 1949, it had outgrown its San Carlos facilities. Russell Varian and Edward Ginzton, whom Terman had steered to the klystron project as a student before the war and who now was a Varian director as well as a Stanford professor, decided to build a branch facility near the university. Early in 1950, Varian asked Stanford to lease land to the firm for the construction of a new research and development facility. The agreement seemed eminently sensible. The firm's research was based on university patents, and Stanford faculty numbered among its important shareholders, directors, and consultants.

This move led Terman, also a Varian director, to give steerage to the university's developing land use master plan. He saw the opportunity to lease only to companies with activities connected to the university's programs, yet the master plan seemed to include other industries, even insurance companies. The second lease to Eastman Kodak for a photo-processing plant mobilized Terman. By 1953, he was on the university's Advisory Committee on Land and Building Development and busy persuading Alf Brandin, the university business manager, to come around to his way of thinking.

Terman pulled his arguments from several directions. He restated that the park could reinforce the university's steeples of excellence in research. He pointed out the large gifts that Stanford had received from Hewlett-Packard, Varian Associates, and Russell Varian. These companies were good donors, with their gifts equaling the university's lease income. He also pointed out the important matching tuition paid by local firms for their employees who participated in the honors cooperative program that enabled outside engineers to sit in on Stanford classes directly or via a television network. "Brandin was quick on the trigger," he recalled. "Very soon thereafter, if you weren't a high-technology company, you had a hell of a time coaxing him to give you a lease."

In 1955, Terman became the university provost and turned to building up Stanford's chemistry department as the foundation for a community of interest in biotechnology. Soon he could see his work bearing fruit. His engineering steeples of excellence made Stanford one of the nation's foremost research universities. Sponsored research climbed from the first ONR project of $225,000 in 1946 to $12 million in 1967. This stimulated even more growth; corporate gifts reached a half-million dollars in 1956 and rose to over $2 million by the year 1965. Meanwhile, Hewlett-Packard became the flagship of Stanford Industrial Park in 1956, and soon, Ampex and Lockheed's new Space and Missile Division moved in. By 1960, over forty firms occupied the 450-acre campus-style park, and Terman was being asked to help mastermind similar feats in Texas and New Jersey.

*   *   *

The Santa Clara Valley, once known by local agricultural boosters as "the Valley of Heart's Delight," had become a growing center for the electronics and aerospace industries. Defense contracts helped drive the boom, with the earliest ones going to Stanford and its spreading community of electronics companies. Development of the NASA-Ames Research Center (opened in 1940) in Sunnyvale, California, also brought contracts and stimulated industry, and San Jose Chamber of Commerce manager Russell Petit spurred the region's growth with an intensive national advertising campaign to attract even more companies. Cashing in on the vacuum tube, radar, microwave technology, Cold War, and space exploration, giant national firms of the tube era of electronics as well as young new entrepreneurs set up shop from Palo Alto south to San Jose. The spreading high-technology community attracted national attention. "Any electrical engineer with ambition ran his resume around the Valley circuit in hope of a response," noted Malone, and seasoned researchers came West as well.

In 1956, William B. Shockley returned to his boyhood home in Palo Alto. As coinventor of the transistor while at Bell Telephone Laboratories in 1947, he had just won the Nobel Prize with John Bardeen and Walter Brattain. Now Shockley would establish the first semiconductor company in the fertile valley. The brightest young engineers in the country responded to his call, and he hired the best. But within two years, his young protégés brewed a rebellion. Disagreeing with Shockley's research directions and fed up with his contemptuous treatment of them, seven began looking for support to set up on their own. With almost no venture capital resources on the West Coast, a New York investment firm finally put them in touch with Fairchild Camera and Instrument Corporation in New Jersey. Fairchild

The first mass-producible monolithic integrated circuit, developed by Fairchild Semi-conductor in 1961.
Source: California History Center Foundation, Cupertino

seemed interested, but was concerned that none of the seven possessed clear management skills. To win the company's support, the seven recruited the only holdout at Shockley who possessed the needed leadership: twenty-seven-year-old Robert Noyce. The now eight men quit en masse before an enraged Shockley. Fairchild Semiconductor opened in Mountain View, California, and Silicon Valley was born.

Numerous problems plagued Fairchild from the start, even though it grew rapidly and introduced the first mass-producible monolithic integrated circuit. Malone described it as "a corporate vocational school" for young engineers. "Here they could screw up without serious repercussions—after all, nobody else knew how the job was done either—and learn from their mistakes." They did, unleashing the centrifugal forces of entrepreneurial creativity that have since characterized Silicon Valley. By 1968, the last of the original eight founders, along with many other Fairchild engineers, had moved on to replicate the founders' experiences in new ventures. Later, Noyce remarked that it came "as a great revelation—and a great motivation, too," that a young engineer or scientist could get venture capital for a new company. At a 1969 conference in Sunnyvale, fewer than two dozen of the four hundred semiconductor engineers present had not worked for Fairchild, and by the early 1970s, former Fairchild employees had started forty-one new semiconductor companies, many of them in the valley.

The only apparent early connection between Fairchild and the Stanford community that Terman had built rested in the firm's early hiring of graduate students as production workers. A closer tie did not come until Marcian "Ted" Hoff joined Noyce's new Intel Corporation after receiving his PhD in electrical engineering at Stanford. In 1969, Hoff developed the microprocessor, the first computer on a chip. Intel's 4004 chip launched another explosive wave through the valley and beyond. By 1975, Intel's third microprocessor, the 8080, became the heart of the first affordable computer, the Altair, produced in Albuquerque, New Mexico.

Popular Electronics' announcement of the Altair in January 1975 brought into existence Silicon Valley's Homebrew Computer Club. Its first meeting in March drew thirty-two people to a home in Menlo Park, near Palo Alto. Within weeks, the meetings were drawing several hundred enthusiasts, and it began meeting in the Stanford Linear Accelerator Center auditorium. According to authors Paul Freiberger and Michael Swaine, the Homebrew Computer Club provided "the intellectual nutrient" in which the valley's microcomputer companies first swam. Homebrewers "thrived in a kind of joyous anarchy" from which solid engineering and technologies for people emerged.

Stephen Wozniak, a young employee at Hewlett-Packard and gifted computer hobbyist, was a Homebrewer from the start. He went regularly, absorbing the home-built machines others brought to the meetings and beginning to feel he could improve on their designs. He bought one of the latest microprocessor chips in conjunction with a San Francisco computer show and built a computer of his own. Wozniak took it to Homebrew, showed it off, and passed out photocopies of his design for the Apple. Meanwhile, his friend Steven Jobs suggested they start a company, and in 1976, Apple Computer was born in the tradition of Hewlett and Packard—in a garage in Cupertino. Eight years later its sales topped $1.5 billion, and Silicon Valley was the envy of the world.

Terman, who had retired from Stanford in 1965, must have been pleased with this latest odyssey in the Santa Clara Valley. After all, it was his vision and life's work that had created the foundation from which it sprang. In 1952, he had observed, "Almost anything that one wishes to do in the world of today is made possible, or is done better, or is helped by electronics. Through its ability to control, to amplify, and to convert between light, sound, and electricity, electronics provides a nervous system for our machine-age civilization." By the time of his death in 1982, his dreams had been more than fulfilled.

For Further Reading

1 Technology in America: An Introduction

Hindle, Brooke, and Steven Lubar. *Engines of Change: The American Industrial Revolution, 1790–1860*. Washington, DC: Smithsonian Institution Press, 1986.

Hughes, Thomas P. *American Genesis: A Century of Technological Enthusiasm, 1870–1970*. New York: Viking, 1989.

Marx, Leo. *The Pilot and the Passenger: Essays on Literature, Technology, and Culture in the United States*. New York: Oxford University Press, 1988.

Morison, Elting E. *From Know-how to Nowhere: The Development of American Technology*. New York: Basic Books, 1974.

Smith, Merritt Roe, ed. *Military Enterprise and Technological Change: Perspectives on the American Experience*. Cambridge, MA: MIT Press, 1985.

2 The Artisan during America's Wooden Age

Bedini, Silvio A. *Thinkers and Tinkers: Early American Men of Science*. New York: Charles Scribners' Sons, 1975.

Hindle, Brooke, ed. *America's Wooden Age: Aspects of Its Early Technology*. Tarrytown, NY: Sleepy Hollow Restorations, 1975.

Hindle, Brooke. *Technology in Early America: Needs and Opportunities for Study*. Chapel Hill, NC: University of North Carolina Press, 1965.

McGaw, Judith A., ed. *Early American Technology: Making and Doing Things from the Colonial Era to 1850*. Chapel Hill: University of North Carolina Press, 1994.

McPhee, John. *The Survival of the Bark Canoe*. New York: Warner Books, 1975.

Sloane, Eric. *Our Vanishing Landscape*. New York: Funk and Wagnalls, 1955.

Sloane, Eric. *A Reverence for Wood*. New York: Funk and Wagnalls, 1965.

3 Thomas Jefferson and a Democratic Technology

Ferguson, Eugene S. *Oliver Evans: Inventive Genius of the American Industrial Revolution*. Greenville, NC: Hagley Museum and Library, 1980.

Hindle, Brooke. *Emulation and Invention*. New York: NYU Press, 1981.

Kasson, John. *Civilizing the Machine: Technology and Republican Values in America*. New York: Grossman Publishers, 1976.

Kouwenhoven, John. *Made in America*. New York: Doubleday, 1948.

Marx, Leo. *The Machine in the Garden: Technology and the Pastoral ideal in America*. New York: Oxford University Press, 1964.

Rigal, Laura. *The American Manufactory: Art, Labor, and the World of Things in the Early Republic*. Princeton, NJ: Princeton University Press, 1998.

4 Benjamin Henry Latrobe and the Transfer of Technology

Jeremy, David J. *Transatlantic Industrial Revolution: The Diffusion of Textile Technologies between Britain and America, 1790–1830*. Cambridge, MA: MIT Press, 1981.

Pursell, Carroll W., Jr. *Early Stationary Steam Engines in America: A Study in the Migration of a Technology*. Washington, DC: Smithsonian Institution Press, 1969.

Stapleton, Darwin H. *The Transfer of Early Industrial Technologies to America*. Philadelphia: American Philosophical Society, 1987.

Tucker, Barbara M. *Samuel Slater and the Origins of the American Textile Industry, 1790–1860*. Ithaca, NY: Cornell University Press, 1984.

5 Eli Whitney and the American System of Manufacturing

Hoke, Donald. *Ingenious Yankees: The Rise of the American System of Manufactures in the Private Sector*. New York: Columbia University Press, 1989.

Hounshell, David A. *From the American System to Mass Production, 1800–1932: The Development of Manufacturing Technology in the United States*. Baltimore: Johns Hopkins University Press, 1984.

Lakwete, Angela. *Inventing the Cotton Gin: Machine and Myth in Antebellum America*. Baltimore: Johns Hopkins University Press, 2003.

Mayr, Otto, and Robert C. Post, eds. *Yankee Enterprise: The Rise of the American System of Manufactures*. Washington, DC: Smithsonian Institution Press, 1981.

McGaw, Judith A. *Most Wonderful Machine: Mechanization and Social Change in Berkshire Paper Making, 1801–1885*. Princeton, NJ: Princeton University Press, 1987.

Noble, David F. *Forces of Production: A Social History of Industrial Automation*. New York: Oxford University Press, 1984.

Rosenberg, N., ed. *The American System of Manufactures*. Edinburgh: Edinburgh University Press, 1969.

6 Thomas P. Jones and the Evolution of Technical Education

Bix, Amy. *Girls Coming to Tech! A History of American Engineering Education for Women*. Cambridge, MA: MIT Press, 2014.

Calvert, Monte A. *The Mechanical Engineer in America, 1830–1910: Professional Cultures in Conflict*. Baltimore: Johns Hopkins University Press, 1967.

Ferguson, Eugene S., ed. *Early Engineering Reminiscences (1815–40) of George Escol Sellers*. Washington, DC: Smithsonian Institution Press, 1965.

Post, Robert Charles. *Physics, Patents, and Politics: A Biography of Charles Grafton Page*. New York: Science History Publications, 1976.

Sinclair, Bruce. *Philadelphia's Philosopher Mechanics: A History of the Franklin Institute, 1824–1865*. Baltimore: Johns Hopkins University Press, 1974.

Williams, Rosalind. *Retooling: A Historian Confronts Technological Change*. Cambridge, MA: MIT Press, 2002.

7 Cyrus Hall McCormick and the Mechanization of Agriculture

Danhof, Clarence H. *Change in Agriculture: The Northern United States, 1820–1870*. Cambridge, MA: Harvard University Press, 1969.

Hutchinson, William T. *Cyrus Hall McCormick*. 2 vols. New York: The Century Co., 1930–1935.

Jellison, Katherine. *Entitled to Power: Farm Women and Technology, 1913–1963*. Chapel Hill: University of North Carolina Press, 1993.

Rossiter, Margaret W. *The Emergence of Agricultural Science: Justus Liebig and the Americans, 1840–1880*. New Haven, CT: Yale University Press, 1975.

Wik, Reynold M. *Steam Power on the American Farm*. Philadelphia: University of Pennsylvania Press, 1953.

Williams, Robert C. *Fordson, Farmall, and Poppin' Johnny: A History of the Farm Tractor and Its Impact on America*. Urbana: University of Illinois Press, 1987.

8   James Buchanan Eads: The Engineer as Entrepreneur

Chandler, Alfred D. *The Visible Hand: Managerial Revolution in American Business.* Cambridge, MA: Belknap Press, 1977.

Hughes, Thomas Parke. *Elmer Speny: Inventor and Engineer.* Baltimore: Johns Hopkins University Press, 1971.

McCartney, Laton. *Friends in High Places. The Bechtel Story: The Most Secret Corporation and How It Engineered the World.* New York: Simon & Schuster, 1988.

McMahon, A. Michal. *The Making of a Profession: A Century of Electrical Engineering in America.* New York: The Institute of Electrical and Electronic Engineers, 1984.

Merritt, Raymond H. *Engineering in American Society, 1850–1875.* Lexington: University of Kentucky Press, 1969.

Sinclair, Bruce. *A Centennial History of the American Society of Mechanical Engineers, 1880–1980.* Toronto: University of Toronto Press, 1980.

9   James B. Francis and the Rise of Scientific Technology

Hunter, Louis C. *A History of Industrial Power in the United States, 1780–1930. Volume One: Waterpower in the Century of the Steam Engine.* Charlottesville: University of Virginia Press, 1979.

Hunter, Louis C. *A History of Industrial Power in the United States, 1780–1930. Volume Two: Steam Power.* Charlottesville: University of Virginia Press, 1985.

McHugh, Jeanne. *Alexander Holley and the Makers of Steel.* Baltimore: Johns Hopkins University Press, 1980.

Noble, David F. *America by Design: Science, Technology, and the Rise of Corporate Capitalism.* New York: Knopf, 1977.

Spence, Clark C. *Mining Engineers and the American West: The LaceBoot Brigade, 1849–1933.* New Haven, CT: Yale University Press, 1970.

10   Alexander Graham Bell and the Conquest of Solitude

Bruce, Robert V. *Bell: Alexander Graham Bell and the Conquest of Solitude.* Boston: Little Brown, 1973.

Fischer, Claude S. *America Calling: A Social History of the Telephone to 1940.* Berkeley: University of California Press, 1992.

Hochfelder, David. *The Telegraph in America, 1832–1920.* Baltimore: Johns Hopkins University Press, 2012.

Smith, George David. *The Anatomy of a Business Strategy: Bell, Western Electric, and the Origins of the American Telephone Industry.* Baltimore: Johns Hopkins University Press, 1985.

Thompson, Robert Luther. *Wiring a Continent: The History of the Telegraph Industry in the United States.* Princeton, NJ: Princeton University Press, 1947.

Wasserman, Neil H. *From Invention to Innovation: Long-Distance Telephone Transmission at the Turn of the Century.* Baltimore: Johns Hopkins University Press, 1985.

11 Thomas Alva Edison and the Rise of Electricity

Friedel, Robert, and Paul Israel. *Edison's Electric Light: Biography of an Invention.* New Brunswick, NJ: Rutgers University Press, 1986.

Josephson, Matthew. *Edison: A Biography.* New York: McGraw-Hill, 1959.

MacLaren, Malcolm. *The Rise of the Electrical Industry during the Nineteenth Century.* Princeton, NJ: Princeton University Press, 1943.

Nye, David E. *Electrifying America: Social Meanings of a New Technology.* Cambridge, MA: MIT Press, 1990.

Passer, Harold C. *The Electrical Manufacturers, 1875–1900: A Study in Competition, Entrepreneurship, Technical Change, and Economic Growth.* Cambridge, MA: Harvard University Press, 1953.

Wachhorst, Wyn. *Thomas Alva Edison: An American Myth.* Cambridge, MA: MIT Press, 1981.

12 Lewis Latimer and the Role of Black Inventors

Fouche, Rayvon. *Black Inventors in the Age of Segregation: Granville T. Woods, Lewis H. Latimer, and Shelby J. Davidson.* Baltimore: Johns Hopkins University Press, 2003.

Friedel, Robert, and Paul Israel. *Edison's Electric Light: Biography of an Invention.* New Brunswick, NJ: Rutgers University Press, 1986.

James, Portia P. *The Real McCoy: African-American Invention and Innovation, 1619–1930.* Washington, DC: Smithsonian Institution Press, 1989.

McMahon, A. Michal. *The Making of a Profession: A Century of Electrical Engineering in America.* New York: The Institute of Electrical and Electronics Engineers, 1984.

Pursell, Carroll, ed. *A Hammer in Their Hands: A Documentary History of Technology and the African-American Experience.* Cambridge, MA: MIT Press, 2005.

Sinclair, Bruce, ed. *Technology and the African-American Experience: Needs and Opportunities for Study.* Cambridge, MA: MIT Press, 2004.

13   George Eastman and the Coming of Industrial Research in America

Hounshell, David A., and John Kenly Smith Jr. *Science and Corporate Strategy: Du Pont R&D, 1902–1980.* Cambridge: Cambridge University Press 1988.

Jenkins, Reese V. *Images and Enterprise: Technology and the American Photographic Industry, 1839–1925.* Baltimore: Johns Hopkins University Press, 1975.

Jewkes, John, David Sawers, and Richard Stillerman. *The Sources of Invention.* London: St. Martin's Press, 1958.

Reich, Leonard S. *The Making of American Industrial Research: Science and Business at GE and Bell, 1876–1926.* Cambridge: Cambridge University Press, 1985.

Wise, George. *Willis R. Whitney, General Electric, and the Origins of U.S. Industrial Research.* New York: Columbia University Press, 1985.

14   Ellen Swallow Richards: Technology and Women

Cowan, Ruth Schwartz. *More Work for Mother: The Ironies of Household Technology from the Open Hearth to the Microwave.* New York: Basic Books, 1983.

Dublin, Thomas. *Women at Work: The Transformation of Work and Community in Lowell, Massachusetts, 1826–1860.* New York: Columbia University Press, 1979.

Hayden, Dolores. *The Grand Domestic Revolution: A History of Feminist Designs for American Homes, Neighborhoods, and Cities.* Cambridge, MA: MIT Press, 1981.

Rothman, Barbara Katz. *Recreating Motherhood: Ideology and Technology in a Patriarchal Society.* New Brunswick, NJ: Rutgers University Press, 1989.

Strasser, Susan. *Never Done: A History of American Housework.* New York: Pantheon Books, 1982.

15   Gifford Pinchot and the American Conservation Movement

Hays, Samuel P. *Beauty, Health, and Permanence: Environmental Politics in the United States, 1955–1985.* Cambridge: Cambridge University Press, 1987.

Hays, Samuel P. *Conservation and the Gospel of Efficiency: The Progressive Conservation Movement, 1890–1920.* Cambridge, MA: Harvard University Press, 1959.

Melosi, Martin V., ed. *Pollution and Reform in American Cities, 1870–1930.* Austin: University of Texas Press, 1980.

Tarr, Joel A., and Gabriel Dupuy, eds. *Technology and the Rise of the Networked City in Europe and America.* Philadelphia: Temple University Press, 1989.

Worster, Donald. *Rivers of Empire: Water Aridity and the Growth of the American West.* New York: Pantheon Books, 1985.

16 Frederick Winslow Taylor and Scientific Management

Aitken, Hugh G. J. *Scientific Management in Action: Taylorism at Watertown Arsenal, 1908–1915*. Cambridge, MA: Harvard University Press, 1960.

Braverman, Harry. *Labor and Monopoly Capital: The Degradation of Work in the Twentieth Century*. New York: Monthly Review Press, 1974.

Copley, Frank Barkley. *Frederick W. Taylor: Father of Scientific Management*. 2 vols. New York: Harper & Bros., 1923.

Gilbreth, Frank B., Jr. *Time Out for Happiness*. New York: Crowell, 1970.

Kakar, Sudhir. *Frederick Taylor: A Study in Personality and Innovation*. Cambridge, MA: MIT Press, 1970.

Nelson, Daniel. *Frederick W. Taylor and the Rise of Scientific Management*. Madison: University of Wisconsin Press, 1980.

17 Henry Ford and the Triumph of the Automobile

Borg, Kevin L. *Auto Mechanics: Technology and Expertise in Twentieth-Century America*. Baltimore: Johns Hopkins University Press, 2002.

Flink, James J. *The Automobile Age*. Cambridge, MA: MIT Press, 1988.

Flink, James J. *The Car Culture*. Cambridge, MA: MIT Press, 1975.

Foster, Mark S. *From Streetcar to Superhighway: American City Planners and Urban Transportation, 1900–1940*. Philadelphia: Temple University Press, 1981.

Leslie, Stuart W. *Boss Kettering: Wizard of General Motors*. New York: Columbia University Press, 1983.

Rose, Mark H. *Interstate: Express Highway Politics, 1941–1956*. Lawrence: Regents Press of Kansas, 1979.

Scharff, Virginia. *Taking the Wheel: Women and the Coming of the Motor Age*. New York: Free Press, 1991.

Seely, Bruce E. *Building the American Highway System: Engineers as Policy Makers*. Philadelphia: Temple University Press, 1987.

18 A. C. Gilbert, Toys, and the Boy Engineer

Cross, Gary. *Kids' Stuff: Toys and the Changing World of American Childhood*. Rev. ed. Cambridge, MA: Harvard University Press, 1999.

Gilbert, A. C., and Marshall McClintock. *The Man Who Lives in Paradise*. New York: Heimburger House, 1954.

McReavy, Anthony. *The Toy Story: The Life and Times of Inventor Frank Hornby*. London: Ebury Press, 2002.

Onion, Rebecca. *Innocent Experiments: Childhood and the Culture of Popular Science in the United States*. Chapel Hill: University of North Carolina Press, 2016.

Pursell, Carroll. *From Playgrounds to PlayStation: The Interaction of Technology and Play*. Baltimore: Johns Hopkins University Press, 2015.

19   Peter L. Jensen and the Amplification of Sound

Aitken, Hugh G. J. *The Continuous Wave: Technology and American Radio, 1900–1932*. Princeton, NJ: Princeton University Press, 1985.

Aitken, Hugh G. J. *Syntony and Spark: The Origins of Radio*. Princeton, NJ: Princeton University Press, 1976.

Czitrom, Daniel J. *Media and the American Mind: From Morse to McLuhan*. Chapel Hill: University of North Carolina Press, 1982.

Douglas, Susan J. *Inventing American Broadcasting, 1899–1922*. Baltimore: Johns Hopkins University Press, 1987.

Horning, Susan Schmidt. *Chasing Sound: Technology, Culture, and the Art of Studio Recording from Edison to the LP*. Baltimore: Johns Hopkins University Press, 2013.

Jensen, Peter L. *The Great Voice*. Richardson, TX: Havilah Press, 1975.

20   Charles A. Lindbergh: His Flight and the American Ideal

Bilstein, Roger E. *Flight in America, 1900–1983: From the Wrights to the Astronauts*. Baltimore: Johns Hopkins University Press, 1984.

Bilstein, Roger E. *Flight Patterns: Trends of Aeronautical Development in the United States, 1918–1929*. Athens: University of Georgia Press, 1983.

Constant, Edward W., II. *The Origins of the Turbojet Revolution*. Baltimore: Johns Hopkins University Press, 1980.

Corn, Joseph J. *The Winged Gospel: America's Romance with Aviation, 1900–1950*. New York: Oxford University Press, 1983.

Rae, John B. *Climb to Greatness: The American Aircraft Industry, 1920–1960*. Cambridge, MA: MIT Press, 1968.

Roland, Alex. *Model Research: The National Advisory Committee for Aeronautics, 1915–1958*. 2 vols. Washington, DC: NASA, 1985.

21 Buster Keaton and Charlie Chaplin: The Silent Film's Response to Technology

Corn, Joseph J., ed. *Imagining Tomorrow: History, Technology, and the American Future.* Cambridge, MA: MIT Press, 1986.

Kasson, John F. *Amusing the Million: Coney Island at the Turn of the Century.* New York: Hill & Wang, 1978.

Marzio, Peter C. *Rube Goldberg: His Life and Work.* New York: Harper & Row, 1973.

Meikle, Jeffrey L. *Twentieth Century Limited: Industrial Design in America, 1925–1939.* Philadelphia: Temple University Press, 1979.

22 Morris L. Cooke and Energy for America

Butti, Ken, and John Perlin. *A Golden Thread: 2500 Years of Solar Architecture and Technology.* Palo Alto, CA: Cheshire Books, 1980.

Christie, Jean. *Morris Llewellyn Cooke: Progressive Engineer.* New York: Garland Publications, 1983.

Hughes, Thomas P. *Networks of Power: Electrification in Western Society, 1880–1930.* Baltimore: Johns Hopkins University Press, 1983.

Melosi, Martin V. *Coping with Abundance: Energy and Environment in Industrial America.* New York: Knopf, 1985.

Williams, James C. *Energy and the Making of Modern California.* Akron, OH: University of Akron Press, 1997.

23 Enrico Fermi and the Development of Nuclear Power

Badash, Lawrence. *Radioactivity in America: Growth and Decay of a Science.* Baltimore: Johns Hopkins University Press, 1979.

Boyer, Paul. *By the Bomb's Early Light: American Thought and Culture at the Dawn of the Atomic Age.* New York: Pantheon, 1985.

Hewlett, Richard G., and Oscar E. Anderson, Jr. *The New World, 1939/1946: Volume I of a History of the United States Atomic Energy Commission.* University Park: Pennsylvania State University Press, 1962.

Hewlett, Richard G., and Francis Duncan. *Nuclear Navy, 1946–1962.* Chicago: University of Chicago Press, 1974.

Nelkin, Dorothy. *Nuclear Power and Its Critics: The Cayuga Lake Controversy.* Ithaca, NY: Cornell University Press, 1971.

York, Herbert F. *The Advisers: Oppenheimer, Teller, and the Superbomb*. San Francisco: W. H. Freeman, 1976.

24   Robert H. Goddard and the Origins of Space Flight

Koppes, Clayton R. *JPL and the American Space Program: A History of the Jet Propulsion Laboratory*. New Haven, CT: Yale University Press, 1982.

Lasby, Clarence G. *Project Paperclip: German Scientists and the Cold War*. New York: Atheneum, 1971.

Lehman, Milton. *This High Man: The Life of Robert H. Goddard*. New York: Farrar, Straus, 1963.

Mazlish, Bruce, ed. *The Railroad and the Space Program: An Exploration in Historical Analogy*. Cambridge, MA: MIT Press, 1965.

McDougall, Walter A. ... *The Heavens and the Earth: A Political History of the Space Age*. New York: Basic Books, 1985.

25   Rachel Carson and the Challenge of Greening Technology

Carson, Rachel. *Silent Spring*. Boston: Riverside Press, 1962.

Lear, Linda. *Rachel Carson: Witness for Nature*. New York: H. Holt, 1997.

Mart, Michelle. *Pesticides, a Love Story: America's Enduring Embrace of Dangerous Chemicals*. Lawrence: University Press of Kansas, 2015.

Murphy, Priscilla Coit. *What a Book Can Do: The Publication and Reception of Silent Spring*. Amherst: University of Massachusetts Press, 2005.

Pursell, Carroll. *Technology in Postwar America: A History*. New York: Columbia University Press, 2007.

Rome, Adam. *The Genius of Earth Day: How a 1970 Teach-in Unexpectedly Made the First Green Generation*. New York: Hill & Wang, 2013.

26   Frederick E. Terman and the Rise of Silicon Valley

Austrian, Geoffrey D. *Herman Hollerith: Forgotten Giant of Information Processing*. New York: Columbia University Press, 1982.

Dickson, David, *The New Politics of Science*. 2nd ed. Chicago: University of Chicago Press, 1988.

Hanson, Dirk. *The New Alchemists: Silicon Valley and the Microelectronics Revolution*. Boston: Little, Brown, 1982.

Kenney, Martin. *Biotechnology: The University-industrial Complex*. New Haven, CT: Yale University Press, 1986.

Kidder, Tracy. *The Soul of a New Machine*. Boston: Little, Brown, 1981.

Markoff, John. *What the Dormouse Said: How the 60s Counterculture Shaped the Personal Computer Industry*. New York: Viking, 2005.

Turkle, Sherry. *The Second Self: Computers and the Human Spirit*. Cambridge, MA: MIT Press, 2005 [1984].

Ullman, Ellen. *Close to the Machine: Technophilia and Its Discontents*. San Francisco: City Lights, 1997.

Young, Jeffrey S. *Steve Jobs: The Journey Is the Reward*. Glenview, IL: Scott Foresman, 1988.

Contributors

Lawrence Badash was a professor of the history of science at the University of California at Santa Barbara. His research dealt primarily with the history of nuclear physics and its applications.

George Basalla is a professor of history at the University of Delaware. He specializes in the social history of technology—particularly how technology relates to culture—and is the author of *The Evolution of Technology* (1989).

Robert V. Bruce was a professor of history at Boston University. In addition to his biography of Alexander Graham Bell, he is the author of *Lincoln and the Tools of War* (1956) and the Pulitzer Prize–winning *The Launching of Modern American Science* (1988).

Jean Christie is a professor of history at Fairleigh Dickinson University. Besides her monograph on Morris L. Cooke, she has published two collections of essays on twentieth-century American history.

Gail Cooper is an associate professor of history at Lehigh University. She is the author of *Air-Conditioning in America: Engineers and the Controlled Environment, 1900–1960* (1998), and is currently working on the adoption of statistical quality control in Japan.

Ruth Schwartz Cowan was a professor of history at the University of Pennsylvania. She is the author of *More Work for Mother* (1983) as well as many articles on the relationship between women and technology.

James J. Flink is a professor of comparative cultures at the University of California at Irvine. He has written three books on the automobile.

Barton C. Hacker is a curator of military history at the Smithsonian Institution. He is the author of *The Dragon's Tail: Radiation Safety in the Manhattan Project, 1942–1946* (1987).

Samuel P. Hays is a professor of history at the University of Pittsburgh. He is the author of *Beauty, Health, and Permanence: Environmental Politics in the United States, 1955–1985* (1987).

Brooke Hindle retired as a senior historian at the National Museum of American History of the Smithsonian Institution. He was the author of many books on American technology, most recently (with Steven Lubar) *Engines of Change: The American Industrial Revolution, 1790–1860* (1986).

Thomas Parke Hughes was a professor of the history of technology at the University of Pennsylvania. His most recent book is *American Genesis: A Century of Technological Enthusiasm, 1870–1970* (1989).

Reese V. Jenkins teaches at Rutgers University and was the director in chief of the Thomas A. Edison Papers. He is the author of *Images and Enterprise: Technology and the American Photographic Industry* (1975).

John A. Kouwenhoven was a professor of English at Barnard College. His books include the classic *Made in America: The Arts in Modern American Civilization* (1948).

Edwin T. Layton Jr. was a professor of history, science, and technology at the University of Minnesota. He is the author of *The Revolt of the Engineers: Social Responsibility and the American Engineering Profession* (1971).

W. David Lewis was Hudson Professor of History and Engineering at Auburn University. He is the author of several books on aviation history, including (with W. P. Newton) *Delta: The History of an Airline* (1979).

Hugo A. Meier taught history at Pennsylvania State University. His writings are centered on concepts of technology in nineteenth-century American social history.

Carroll Pursell is an Adeline Barry Davee Distinguished Professor of History (emeritus) at Case Western Reserve University and an adjunct professor of history at the Australian National University.

Adam Rome is a professor of history at the University at Buffalo, SUNY. He is the author of *The Bulldozer in the Countryside* (2001) and *The Genius of Earth Day: How a 1970 Teach-in Unexpectedly Made the First Green Generation* (2013).

Bruce Sinclair was the Melvin Kranzberg Professor of History at the Georgia Institute of Technology. His most recent book is *Technology and the African-American Experience: Needs and Opportunities for Study* (2004).

Merritt Roe Smith is a professor of the history of technology in MIT's Program on Science, Technology, and Society. His book *Harpers Ferry Armory and the New Technology* (1977) won the Frederick Jackson Turner Award from the Organization of American Historians.

Darwin H. Stapleton is the editor of *The Engineering Drawings of Benjamin Henry Latrobe* (1980) and was the associate editor of the three-volume *Correspondence and Miscellaneous Writings of Benjamin Henry Latrobe* (1984–88). He is the executive director emeritus of the Rockefeller Archive Center.

John William Ward was president of Amherst College and directed the American Council of Learned Societies.

James C. Williams is a visiting research professor at Stetson University.

Index